D0405613

TEDBooks

Follow Your Gut
The Enormous Impact
of Tiny Microbes

ROB KNIGHT

WITH BRENDAN BUHLER

TED Books

Simon & Schuster

New York   London   Toronto   Sydney   New Delhi

 **TED**Books

Simon & Schuster
1230 Avenue of the Americas
New York, NY 10020

First TED Books hardcover edition April 2015

For information about special discounts for bulk
purchases, please contact Simon & Schuster Special Sales
at 1-866-506-1949 or business@simonandschuster.com.

For information on licensing the TED Talk that accompanies
this book, or other content partnerships with TED, please
contact TEDBooks@TED.com.

Interior design by MGMT. design
Illustrations by Olivia de Salve Villedieu

Manufactured in the United States of America

10 9 8 7 6 5 4 3 2

Library of Congress Cataloging-in-Publication Data is available.

ISBN 978-1-4767-8474-8
ISBN 978-1-4767-8475-5 (ebook)

To my parents, Allison and John, for their genes, ideas, and microbes.

CONTENTS

Follow Your Gut

You, we know: human, bipedal, noble in reason, infinite in faculties, heir to all creation, has never read a single end-user license agreement—just checks the box. Now meet the rest of you: the trillions of tiny creatures living in your eyes, your ears, and inside the magnificent mansion that is your gut. This microscopic world within our bodies holds the potential to redefine how we understand disease, our health, and ourselves.

Thanks to new technologies, many of them developed only within the past few years, scientists today know more about the microscopic life-forms inside us than ever before. And what we're learning astonishes. These single-celled organisms—microbes—are not only more numerous than we thought, inhabiting in enormous numbers almost every nook and cranny of the body, but they're also more important than we ever imagined, playing a role in nearly all aspects of our health, even in our personality.

The collection of microscopic critters that make their home in and on us is called the human microbiota, and their genes are called the human microbiome. And like many scientific breakthroughs, the emerging facts about this tiny world serve as a rebuke to our egos. Astronomy has told us that our planet was not the center of the universe, and evolution says humans are merely one animal among many. The charting of the human

UP TO
10x
MORE

TRILLION CELLS

110
100
90
80
70
60
50
40
30
20
10

HUMAN CELLS   MICROBE CELLS

microbiome teaches us that even within our own bodies, we're drowned out by a chorus of independent (and *inter*-dependent) life-forms with their own goals and agendas.

Just how many microbes are there within us? You are made up of about ten trillion human cells—but there are about a hundred trillion microbial cells in and on your body.[1] Which means: you are mostly not you.

But we are not, as we have thought, merely the unlucky hosts to the occasional bad bug that gives us an infection. In fact, we live in balance with a whole community of microbes all the time. Far from being inert passengers, these little organisms play essential roles in the most fundamental processes of our lives, including digestion, immune responses, and even behavior.

Our inner community of microbes is actually more like a collection of different communities. Different sets of species inhabit different parts of the body, where they play specialized roles. The microbes that live in your mouth are distinct from those residing on your skin or in your gut. We are not individuals; we are ecosystems.

Our diversity of microbes can even help explain certain corporeal quirks that we've long just chalked up to luck, good or bad. For instance, why do some of us seem to taste better to mosquitoes? The little fiends seldom bite me, but my partner, Amanda, attracts them in swarms. It turns out that some of us really *are* more appetizing to mosquitoes than others. And one important reason for our variable delectability is the different microbial communities we harbor on our skin. (More on this in chapter 1.)

And it doesn't end there: there is extraordinary varia-tion in the microbes that live in and on each of us. You've probably heard that we're all pretty much the same in terms of our human DNA: that, in terms of your human DNA, you're 99.99 percent identical to the person sitting next to you. But that's not true of your gut microbes. You might only share 10 percent with the person next to you.

These differences may account for an enormous range of variations between us, from weight to allergies; from our likelihood of getting sick to our level of anxiety. We are only just beginning to map—and to understand—this vast microscopic world, but the implications of our findings are stunning.

The incredible diversity of the microbial world is made all the more mind-blowing by the fact that, until about forty years ago, we had no idea just how many single-celled organisms there were, or how many kinds. Until then, our basic ideas about categorizing the world's living things came from Charles Darwin's *On the Origin of Species*, published in 1859.[2] Darwin sketched out an evolutionary tree that grouped all living things by their shared physical traits—short-beaked finches, long-beaked finches, and the like—and that became our basis for sorting species.

This traditional picture of life was based on what people could see in the world around them or through microscopes: larger living things were classified as plants, animals, and fungi. The remaining single-celled organisms were lumped into two basic categories: protists and bacteria. We were right about the plants, animals, and fungi. But our picture of single-celled organisms was completely wrong.

In 1977, American microbiologists Carl Woese and George E. Fox mapped the tree of life by comparing

THE TREE OF LIFE

HERE LIES EVERYTHING YOU HAVE
HERETOFORE CONSIDERED LIFE

EUKARYOTES

plants

fungi

animals

Ciliates

flagellates

diplomonads

BACTERIA

cyanobacteria

proteobacteria

bacteroidetes

firmicutes

actinobacteria

verrucomicrobia

extreme
halophiles   methanogens   extreme thermophiles

ARCHAEA

SURPRISE! HEREWITH IS THE ACTUAL KNOWN DIVERSITY
OF LIFE (SO FAR)

life-forms at the cellular level, using ribosomal RNA, a relative of DNA that's housed in every cell and used in making proteins. The result was startling.[3] Woese and Fox revealed that single-celled organisms are more diverse than all of the plants and animals combined. As it turns out, animals, plants, and fungi; every human, jellyfish, and dung beetle; every strand of kelp, patch of moss, and soaring redwood; and every lichen and mushroom—all the life we can see with our eyes—amount to three short twigs at the end of one branch on the tree of life. The single-celled organisms—bacteria, archaea (which were discovered by Woese and Fox), yeasts, and others—dominate.

In just the last few years, we've taken amazing leaps forward in our understanding of the microscopic life within us. New techniques—including improvements in DNA sequencing—combined with an explosion of computing power, have been key here. Now, through a process called next-generation sequencing, we can collect cell samples from different parts of the body, rapidly analyze the microbial DNA they contain, and combine information from samples across the body to identify the thousands of species of microbes that call us home. We're finding bacteria, archaea, yeasts, and other single-celled organisms (such as eukaryotes) that collectively have genomes—the genetic recipes that define them—longer than our own.

New computer algorithms, in turn, are making it much easier to interpret all this genetic information. Specifically, we can now create a map of our microbes to compare communities in different parts of the body, and to compare different people's communities to one another.

Much of our growing knowledge comes from the Human Microbiome Project. The $170 million research effort, sponsored by the US National Institutes of Health (NIH), has supported more than two hundred scientists who, so far, have analyzed at least 4.5 terabytes—that's 4.5 trillion bytes—of DNA data. And this is only a start; other international efforts, like MetaHIT (a European consortium), are adding and analyzing more data all the time.

The cost of this analysis is dropping quickly, enabling many more individuals to obtain a census of the diverse life within them. About a decade ago, if you wanted to know what made up your microbiome, you'd need $100 million. Now obtaining that same information costs about $100—so cheap that it might soon become a routine medical procedure ordered by your doctor.

Why would your doctor want to know about your microbiome? Because new research is emerging that suggests previously unknown links between our microbes and numerous diseases, including obesity, arthritis, autism, and depression. And as we begin to shine a light on those links, we're seeing glimmers of future treatments. Almost anything you can imagine has an effect on the microbiome: medicine, diet, whether you're the oldest child, or how many sexual partners you have. As you'll read in the pages that follow, we're discovering that microbes are deeply integrated into almost all aspects of our lives. Indeed, microbes are redefining what it means to be human.

1   The Body Microbial

Just how much microscopic life dwells inside you?

If we're going by weight, the average adult is carrying about three pounds of microbes. This makes your microbiome one of the largest organs in your body—roughly the weight of your brain and a little lighter than your liver.

We've already learned that, in terms of sheer numbers of cells, the microbial cells in our bodies outnumber the human cells by up to ten to one. What happens if we measure by DNA? In that case, each of us of us has about twenty thousand human genes. But we're carrying some two million to twenty million microbial genes. Which means that, genetically speaking, we're at least 99 percent microbe.

If you want a salve for human dignity, think of this as a matter of complexity. Every human cell contains many more genes than a microbial cell. But you contain so many microbes that all their various genes add up to more than your own.

The organisms that live within and upon us are many and varied. Most, but not all of them, are single-celled organisms. They come from all three main branches of the tree of life. You might find in your gut members of the archaea, single-celled organisms that make do without nuclei; the most common of these are the methanogens, creatures that exist without oxygen, help digest our food,

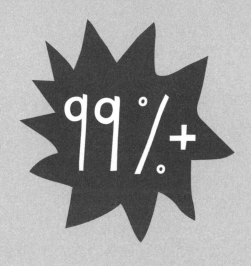

99%+

OF THE GENES IN
OUR BODIES COME
FROM OUR MICROBES

and excrete methane gas. (Cows have them, too.) Then there are eukaryotes, such as the fungi of athlete's foot, and the yeasts that colonize the vagina and sometimes our gut. And most dominant of all, there are our bacteria, like *Escherichia coli*, which we think of mostly as an illness to be caught from underwashed spinach but that actually exists in harmless and helpful versions within most human intestines.

And every day, with the help of new technology, we discover that these creatures are still more diverse than we knew. It is almost as though we had been trawling the ocean with a very wide-meshed net and concluded that marine life consisted entirely of whales and giant squid. We've discovered that there's so much more out there. For instance, you might guess that any two bacteria in your gut, feeding on your latest sandwich, are pretty similar—as similar as, say, an anchovy and a sardine. In fact, their differences are more like a sea cucumber and a great white shark: two creatures with radically different behaviors, nutrient sources, and ecological roles.

So where are all of our microbes, and what are they doing? Let's take a tour of our bodies and find out.

Skin

Napoléon I, returning from a campaign, supposedly sent a message ahead to the empress Joséphine: "I will return to Paris tomorrow evening. Don't wash." He preferred his beloved's scent, and a lot of it. But why is it that, when we are disarmed of our soaps, antiperspirants, powders, and perfumes, we stink so? Largely because of microbes that feast on our secretions and make them yet smellier.

Scientists are still sniffing out what productive purpose the creatures living on our largest organ, our skin, might serve, but this much is certain: they contribute to our body odor, including the scents that attract mosquitoes.[1] As noted earlier, mosquitoes *do* prefer the smell of some individuals over others, and microbes are responsible. These microbes metabolize the chemicals our skin produces into different volatile organic compounds that the mosquitoes like or dislike. Different species of mosquitoes favor different parts of our bodies. For *Anopheles gambiae*, one of the main mosquitos that transmit malaria, the alluring odors aren't coming from our armpits but from our hands and feet. This raises the intriguing possibility that antibiotics rubbed on the hands and feet might ward off attacks from this particular mosquito because by killing the microbes, you'd kill the smell.

Like all our microbes, those on our skin don't necessarily exist for our benefit. But as benign inhabitants, they actually help us a great deal: simply by taking up residence upon us, they make it harder for other, nastier microbes to infect us. Different parts of the skin have different microbes on them, and the diversity—the number of different kinds of microbes—isn't necessarily linked to the number of individual microbes you have in a particular location. Often it's the reverse. In American terms, imagine if Vermont (population, six hundred thousand) had the ethnic diversity of Los Angeles County (population, ten million), and Los Angeles was as monochromatic as Vermont. Your armpits and your forehead have a lot of microbes but relatively few species; in contrast, the palm of the hand and the forearm are sparse microbial habitats,

but many species accumulate there.[2] Women tend to have more diverse microbial communities on their hands than men do, and these differences survive hand washing, suggesting that they may stem from biological differences, although the cause is still unknown.[3]

We've also found that the microbes living on your left hand are different from the microbes on your right. For all of the hand-wringing, knuckle popping, and touching of the same surfaces that our hands do, each develops distinct microbial communities. This inspired Noah Fierer, a professor of ecology and evolutionary biology at the University of Colorado, in Boulder, and me to try to reproduce one of the most famous findings in large-scale biology. British biologist and anthropologist Alfred Russel Wallace and others developed an elaborate theory of biogeography to help explain the dispersal of organisms among islands, and the relationship between species diversity and land area.[4] Wallace, Darwin's contemporary and codiscoverer of natural selection, discovered a split running through present-day Indonesia and Malaysia that separates the Asian fauna (monkeys, rhinoceroses) from the Australian fauna (cockatoos, kangaroos). Fierer and I wanted to know if we could find the same "Wallace line" between the letters G and H on computer keyboards, with distinct populations from its users' left and right hands colonizing each half of the device. We also wondered if the space bar had more kinds of microbes simply because it's larger than the other keys.

Our results suggested a sort of Wallace line, but we were astonished to find something much more remarkable: each fingertip and its corresponding key had essentially the

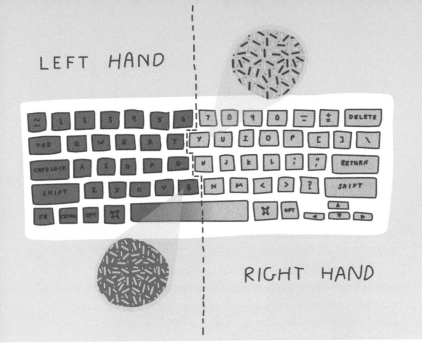

LEFT HAND

RIGHT HAND

same microbial community. We could also match up some-one's computer mouse to the palm of his or her hand with more than 90 percent accuracy.[5] The microbes on your hands are very distinct from other people's—on average, at least 85 percent different in terms of species diversity— which means that you have a microbial fingerprint.

We have taken our research further, performing experiments to understand how many times you have to touch an object to leave these detectable microbial traces. The science is still too preliminary to stand up in a court of law. But TV crime dramas employ, shall we say, slightly lighter standards of evidence, so, shortly after we published the first paper on the topic, *CSI: Miami* aired an episode that used microbial forensics as its premise.[6]

Meanwhile, David Carter, a forensic microbiologist,

recently moved from Nebraska to Hawaii, where he is setting up a body farm. A body farm, you ask? Forensic scientists have to figure out how long corpses they find have been dead. Inside a forensic facility, donated bodies are laid out in different death scenes[7] and then examined every so often to see how they are decomposing. There's a remarkable microbial succession. Just like bare rock is colonized first by lichens, and then, sequentially, mosses, grasses, weeds, shrubs, and, finally, trees, the process of decay follows a very predictable pattern.

Jessica Metcalf, a postdoctoral researcher in my lab at the University of Colorado Boulder, has built her own miniature body farm using forty dead mice. (The mice were killed as a by-product of other experiments aimed at discovering cures for heart disease and cancer.) She found she could estimate when the mice had died within a three-day window, which is about as accurate as current insect-based methods[8] for dating corpses. Why use microbes, then? Insects have to find the corpse, whereas the microbes are right there with you all along, which could make them useful in insect-free crime scenes.

Nose and Lungs

Moving along on our tour, let's look up your nose. The human nostril harbors its own distinct microbes. Among them is *Staphylococcus aureus*, the bacterium that causes staph infections in hospitals. Healthy people, it seems, are often home to what we consider dangerous microbes. What we think might be going on here is that the other bacteria you have in your nose may prevent *S. aureus* from gaining a foothold, or, rather, a nose hold. Another interesting finding is that our environments greatly influence the types of microbes that gather in our noses. And children who have more diverse kinds of bacteria in their nose early on, such as those who live on or near farms, are less likely to develop asthma and allergies later in life.[9] It turns out that playing in the dirt can be good for you.

Down in your lungs, we usually find only dead bacteria.[10] The lungs' air-exposed surfaces contain a cocktail of antimicrobial peptides: tiny proteins that kill bacteria as soon as they land. In sick people, though, such as those with cystic fibrosis or human immunodeficiency virus (HIV), you'll sometimes find harmful microbes that contribute to pulmonary disease.[11]

Whether your throat has its own distinct microbiome, or just has microbes from the mouth passing through, is still a matter of scientific debate.[12] However, we can say that the microbes in the throats of smokers appear to be different from those of nonsmokers, perhaps showing that smoking is harmful not just to us but also to some of the creatures that live within us.[13]

Mouth and Stomach

You've probably heard only about the bad bacteria in your mouth—the ones that can cause gum disease and tooth decay. One bad bug is called *Streptococcus mutans*, a creature that likes to eat our teeth. It seems to have evolved along with human agriculture,[14] which made our diets much richer in carbohydrates, especially sugars. Just as we have inadvertently domesticated rats to eat food out of the garbage, bacterial vermin have been domesticated to live in our bodies. Fortunately, most of the domesticated bacteria in our mouths are beneficial, forming biofilms that keep out the bad bacteria. Our mouth microbes may even help regulate our blood pressure by relaxing our arteries with a compound they help produce called nitric oxide (a chemical relative of the nitrous oxide you've experienced in the dentist's chair).

Another species, called *Fusobacterium nucleatum*, is normally found in healthy mouths but can also contribute to periodontal disease.[15] *F. nucleatum* is interesting because it has been observed within the tumors of people with colon cancer.[16] We don't know yet whether this association is cause or effect: *F. nucleatum* might produce the disease, or it might simply be responding to the environment where the tumor lives. The microbes in your mouth are also quite diverse. Even different sides of the same tooth can harbor their own microbial communities, which could be influenced by many factors, including oxygen exposure and chewing patterns.

In the stomach, we find a highly acidic environment, like a car battery, where only a few kinds of microbes survive. But those microbes can be very important. One in particular, *Helicobacter pylori* (or *H. pylori*), has lived with us for so long that we can tell which human populations are closely related—and with whom they came in contact as they migrated—by looking at the particular strains of *H. pylori* they harbor.[17]

H. pylori play a key role in ulcers, those sores in the stomach or small intestine where the protective mucous lining has been worn away and gastric acid gnaws at the body's tissue. Symptoms start with bad breath and burning stomach pain, and escalate to nausea and bleeding out of both ends. For years, doctors blamed ulcers on stress and diet, advising patients to relax and cut out spicy foods, alcohol, and coffee. Milk and antacids were recommended. Patients experienced some relief but rarely recovered fully.

Then in the 1980s, Australian physicians Barry Marshall and J. Robin Warren showed that most ulcers are actually caused by *H. pylori* infections and can be treated with antibiotics or chemicals such as bismuth, which target the bacteria. In fact, Marshall was so convinced of this that he drank a culture of *H. pylori*, gaining himself curable gastritis and a Nobel Prize, the latter of which he shared with Warren.

And yet today we are learning that something like half of the entire human population carries *H. pylori*. So why don't they all have ulcers? It seems that *H. pylori* is only one of many risk factors for ulcers: necessary but not sufficient. *H. pylori*, along with a great many bacteria we associate with disease, turns out to be something that many people carry without complaint. One of the challenges and promises of microbiome science is figuring out how and why these microbes sometimes turn on us.

Intestines

Next, we come to the intestines. We believe this to be the largest and most important microbial community in the body. If you're a microbe living on a human, this is the main act. Here is the great mansion of our gut, some twenty to thirty feet long and full of nooks and crannies. It's good living for microbes: warm, plenty to eat, plenty to drink, and a convenient sewer system. With their huge microbe populations and wealth of available energy, our intestines are both bustling New York City and oil-rich Saudi Arabia.

The small intestine is where most of the nutrients from your food are absorbed into the bloodstream. The large intestine is where water is absorbed, and helpful microbes ferment the fiber in your diet, which has passed undigested from the small intestine. This releases even more energy for you to harness. And because they work alongside our digestive system, intestinal microbes are in many ways the gatekeepers of our metabolism. They have the potential to influence what we can eat, how many calories we derive from it, what nutrients and toxins we're exposed to, and how drugs affect us.

The other great fact about this highly important collection of microbial communities, scientifically speaking, is that it's so easy to obtain samples. The microbes just slough off and pass right on out, living and dead, usually following coffee in the morning. Your feces contain microbes mostly from the distal large intestine, which is nearer to the end of the line.[18] While there are differences between the microbes in your small and large intestines, in general, this variation is tiny when compared with that found between individual people.[19] This means your poop is a good readout of the microbes unique to your gut.

Of course, in some ways, the microbial picture we derive from poop is a distorted one. For instance, *E. coli* has made lots of headlines as a seemingly ominous bacterium that occasionally makes its way into your food at restaurants with less-than-clean kitchens—but it's not necessarily menacing by itself. We hear about it only because it's easy to find in feces. (If you find *E. coli* in your meat or vegetables, it means they've been contaminated by fecal matter.) In fact, *E. coli* is not actually a major player in the gut, making up less than one cell in ten thousand of most healthy adults.[20] Its fame owes to the fact that it's basically a weed, the dandelion of bacteria, and it grows really well in a petri dish. The same is true of other bacteria that for years have played outsized roles in our understanding of our microbiome: we know about them because they're easy to grow in a lab.

Most of the microbes in our gut are much more fickle, and we don't know yet how to cultivate them in vitro (that is: in the lab). These gut microbes, mainly from two major groups of bacteria called the Firmicutes and the Bacteroidetes,[21] are important for digesting food and for metabolizing drugs, but they have also been linked to a range of diseases including obesity,[22] inflammatory bowel disease, colon cancer, heart disease,[23] multiple sclerosis,[24] and autism.[25] This is why techniques such as next-generation DNA sequencing have been such a revolution. We're finally able to look at what until now has been invisible.

Genitals

First, a confession of ignorance: we don't yet know a great deal about the microbes residing on and inside the penis. For a field founded by the semen-peeping Dutch scientist Antonie van Leeuwenhoek (see pages 59–61), modern microbiology hasn't taken as close a look at male genitalia. However, some progress is being made.

I have a colleague (who shall remain anonymous, lest he be hounded by some cable TV reporter) who performs very important research on the sexually transmitted disease risks among teenagers. A small part of his work looks at the microbiome in and on the penises of teenage boys. For this, he needs samples—gathered both at regular intervals and after the adolescents have a sexual event. So when this colleague gets a call from one of these subjects, he sets off in his white-panel research van—wearing his long hair and standard wardrobe of a leather jacket and gold chain around his neck—to collect samples from your teenager's penis. Sure, it's all for science. But boy, those are some level-headed parents signing the consent forms. Anyway, possibly due to the giggle factor, not a lot of research has been done in this area, meaning this colleague's work will be an (if not *the*) important portrait of the penile microbiome in health and sickness.

The vagina, however, has been studied extensively. In healthy adult women of European ancestry, the vagina is usually dominated by just a few species of *Lactobacillus*. No, those are not the same *Lactobacillus* species you find in yogurt, but they are closely related and also produce lactic acid, keeping the vagina acidic. Work performed by Jacques Ravel, a professor of microbiology and immunology at the University of Maryland, shows that the species dominating a particular woman's vaginal community can vary over time, including during her menstrual cycle, when iron-metabolizing bacteria called *Deferribacter* feed on the blood.[26] A woman's vaginal bacteria can even change when she starts sleeping with a new sexual partner.

Until recently, most of the research on vaginal bacteria has focused on sexually transmitted infections. Scientists have investigated the role of vaginal microbes in a disease called bacterial vaginosis, and they've examined whether vaginal microbes might help or hinder the transmission of various sexually transmitted infections, including HIV.

But it turns out that not all healthy vaginal microbiomes look alike. New research suggests that different populations—Hispanics, African Americans, Caucasians, and Asians, among others—have very different healthy vaginal microbial communities. And as we'll see, in some respects, vaginal microbes define our destiny.

2 How We Get Our Microbiome

When you are a parent, you want the best for your child. When you are a scientist, sometimes you have a very specific idea, based on observational data and statistical analysis, about what's best. And when you're a scientist like me, who studies the role of the microscopic life inside us from birth onward, these ideas can sometimes play out in, shall we say, unusual ways.

When my partner, Amanda, and I were expecting our first child, we had a very detailed birth plan, complete with a doula (a birth assistant—sometimes it's really nice to have someone who's actually on your side, not your health insurance company's side). But kids, even before birth, have no great respect for plans. On November 2, 2011, the Human Microbiome Project writing team, which I was part of, had finally submitted the two main papers describing its results to *Nature*, a leading scientific journal. It had been a long struggle and had taken a toll on Amanda and me. We were owed a little celebration. But Amanda was still pregnant, so I had to drink for both of us—or perhaps for all three. Whatever. Our daughter wasn't due for another three weeks. The vast amount of baby stuff still to be done could wait until the morning.

Around midnight, we were retiring to bed when Amanda suddenly got a weird look on her face. Reaching down to touch the carpet below her feet, she said, "I think my water just broke." She called the hospital, and they told us to come in. We hastily dressed, jumped in the car, and Amanda drove to the hospital, just two and a half miles from our house. Our obstetrician confirmed that Amanda's water had, in fact, broken, and the baby was coming soon, three weeks earlier than expected. Okay, we said, we'll just go back and get our baby stuff—onesies, blankets, bottles—which we had acquired but not yet packed. They informed us that Amanda would not be leaving the hospital until the baby came out.

This presented a dilemma: I was in no condition to drive, although I felt that I was sobering up very rapidly. I called a taxi, but the driver got lost trying to find the hospital (our area is not exactly New York City in terms of cab access), and he wasn't close after an hour. So I told him to forget it, and walked home in the snow with the detailed list of items we'd need. I managed to stuff everything on the list into all three backpacks we owned and trekked back to the hospital.

Everything was going okay. Or so it seemed. But after twenty-four hours in the hospital, the physicians were increasingly concerned. They told us that our baby was in fetal distress. We consulted with our doula, and she agreed that this was the point where we should really stop hoping that nature would take its course and rely on modern medicine. Our daughter was born via an unplanned cesarean section, and I was holding her twenty minutes later. But today's medical technology doesn't supply everything. When it came to her microbes, we took matters into our own hands and swabbed her with samples from Amanda's vagina. Our baby needed those microbes.

When we tell people this story, they typically have three questions. To the first, we're telling you this because we're practicing for when we will tell it to our daughter's prom date.

To the second, how we did it, well, there is no established procedure, but we used sterile cotton swabs (medical-grade Q-tips, basically) to take samples from the vagina, which we then transferred to various parts our newborn: the skin, the ears, the mouth—all the places microbes would have ended up naturally had she passed through the birth canal.

But as for your third excellent question—why we thought this was a good idea in the first place—that's going to take some time to explain.

You get your first microbes from your mom, as you pass through the birth canal. And there's evidence that, before you are even born, your mother's microbiome is preparing for you. During pregnancy, particular kinds of *Lactobacillus*[1] begin to dominate a woman's vagina. The population in her gut shifts toward microbes more efficient at extracting energy from what she eats. This population is also, unfortunately, more likely to cause inflammation of the gut, especially during the third trimester, which is a complex phenomenon that contributes to, among other problems, diarrhea and cramps.

How do we know there's a change in a woman's microbiome during pregnancy? The answer involves syringes full of feces and our much-put-upon assistant, the laboratory mouse. An international team of scientists from the United States, Finland, and Sweden transferred stool from pregnant women into mice raised in a sterile bubble, which left them without any microbes of their own. The mice were divided into two groups. One group received fecal matter from women in their first trimester, while the other mice were given samples from the third trimester. Both groups of mice were fed identical diets. The third-trimester mice, however, gained more weight and had a range of other metabolic and immunological characteristics seen often during pregnancy.[2]

By transplanting the microbes into mice, we can investigate whether these population changes are responses to pregnancy, or whether microbes are the catalysts. The microbial communities in the guts of pregnant women might be changing so that moms can extract more energy or nutrients from their diet in order to pass them on to their baby. It's also possible that those gut microbes are preparing to pass themselves on to the fetus. We know this happens in animals with specialized diets, such as koalas, which need to digest eucalyptus leaves, and vampire bats, which need to digest blood.

It's still not entirely clear whether we have any microbes when we're still in the uterus. Reports have linked microbes in the amniotic fluid or in the placenta to preterm birth.[3] But these initial findings have not been widely reproduced. The current thinking is that healthy fetuses probably don't have any bacteria, although, as with everything in science, this may be subject to revision as new data accumulate.

Your first microbes probably come during birth. You gain them while passing through your mother's birth canal, which is lined with vaginal bacteria. Although different women can have rather different vaginal microbial communities, during pregnancy their microbial communities all move into the same state.[4] And this makes a lot of sense if, as we believe, those microbes evolved to coat the baby with a protective layer against the world. It's a bit like one of those cartoons in which a new child is welcomed into the world by butterflies and songbirds flitting about it—if only the birds and butterflies were fuzzy-looking microbes shoved into the kid by the slimy snootful.

Now, let's assume a baby's first microbes come from the mother's birth canal and vagina. What happens if you're not born that way? Cesarean section births are on the rise in many countries[5] either because of increasing rates of medical complications or simply because they are easier to schedule.

Maria Gloria Dominguez-Bello, a researcher at New York University's Langone Medical Center, studies the microbiome of human infants. I worked with Dr. Dominguez-Bello to show that unlike adults, who contain many distinct microbial ecosystems, the micro-biomes in newborn babies appear more or less the same. If they're delivered vaginally, their microbes look like their mothers' vaginal communities; if they're delivered by cesarean, their microbes look like those found on adult skin, a completely different community.[6] Cesarean births are associated with higher rates of a broad range of diseases associated with the microbiome and/or the immune system, including asthma[7] and, although different studies conflict at the moment, possibly obesity,[8] food allergies,[9] and atopic disease (a kind of skin rash).[10] But don't panic if you or your child was born by cesarean.

IN MICROBES AS IN LIFE...

WE SHARE SOME THINGS
BUT NOT ALL.

MICROBES FOUND IN BABY BOY'S FECES

- ■ ADULT ORAL MICROBES
- ■ ADULT VAGINAL MICROBES
- □ ADULT SKIN MICROBES
- ■ ADULT FECAL MICROBES
- ● BABY BOY'S FECAL MICROBES

AT BIRTH, THE BABY'S MICROBES RESEMBLE VAGINAL MICROBES...

START

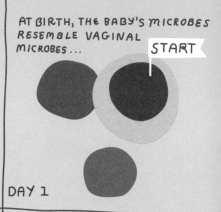

DAY 1

AS HE GROWS...

HIS MICROBES CHANGE...

THE BOY IS GIVEN ANTIBIOTICS

WHICH CAUSE HIS MICROBES TO REGRESS...

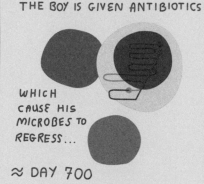

≈ DAY 700

THE BABY BOY'S MICROBES RECOVER...

KOENIG ET AL, 2011

THE BOY'S FECAL MICROBES FINALLY MATCH THOSE OF ADULTS.

END

DAY 838

The most likely outcome is that you'll be just fine. We're talking about increases in relatively small risks.

And yet it makes sense that missing an exposure to a community of microbes we're adapted to might lead to health problems. Until the past century or so, every human who survived to adulthood had been delivered through the birth canal and coated with its community of microbes. This is why when our own child was born by unplanned cesarean, we coated her with the vaginal microbes that she would have received naturally. In the absence of any official guidance on how to do this, we went with the swabs.

We don't know yet if this has had an effect on our daughter—you can't do meaningful statistics with a sample size of one baby. But my lab is conducting a pilot study with Dr. Dominguez-Bello to test whether it has an effect more generally. As of this writing, we have been able to confirm that vaginal and cesarean babies have different microbiomes immediately following birth (as has a Canadian group conducting a similar study [11]), although we don't have enough information yet to determine if or how it impacts health later in life.

It's also hard to tease out the effects of a cesarean versus vaginal birth, because once we're born, our microbiomes quickly become truly complex. At the moment of birth, all of us who arrived here via the vagina have very similar microbiomes. But by the time we're adults, the differences between us are huge.

If we can be so different from one another, you might wonder who we are most similar to. Is it people who eat the same foods as we do? Is it family members we share a home with? Is it residents of our city or continent? It turns out that all of these factors influence our microbiomes, and we're just starting to discover that some are more important than others.

One of the most profound periods in the development of our microbiota (those are our microbes themselves, whereas *microbiomes* refers to their genes) occurs while we are infants. Cornell University's microbiology professor Ruth Ley and my lab performed a study that tracked a single child from the very first stool he produced on through the first 838 days of life.[12] We found that a vaginally delivered baby boy's gut microbiome starts off looking like he has an adult woman's vaginal community in his stool (which is to be expected from his delivery mode) and eventually develops into a normal-looking adult microbiome. But in between those two points in time, what variation!

Day to day, the differences between his stool communities are much larger than the differences between the fecal microbes of two healthy people. In some cases, the difference in his microbes one week to the next is greater than the difference we saw in the 250 adults we tracked in a related project. Microbially speaking, the boy starts off looking more or less like a bear (bears have very simple guts due to their meat-rich diet) and ends up looking like a monkey. One feature that's immediately apparent is a period when the child receives antibiotics for an ear infection, which makes him look not just like a different person but also almost like a different species. However, within a few weeks of recovering, he goes on to resemble the adult microbial state. It raises questions about how frequently we dose our children, and ourselves, with antibiotics.

YOU ARE WHAT YOU EAT

BACTEROIDES   PREVOTELLA

Our diets also help shape our microbiome, even very early on. There are substantial changes associated with breast feeding versus formula feeding. A breast-fed infant is exposed to special microbes found in breast milk, as well as to special sugars in breast milk that promote the growth of beneficial microbes. Our microbiomes then evolve further when we're introduced to solid food. At this stage, around six months old, short-term dietary interventions tend to have very little impact on our microbiome compared to the microbial differences among different people. However, in the long term, you are what you eat: diet over a period of a year has one of the largest effects we've seen on the gut microbiome, adjusting the balance of two major groups of bacteria that digest protein and dietary fiber.[13]

These two categories of intestinal bacteria also account for a little-celebrated aspect of global diversity: gut microbiomes. That's right: along with distinct languages and cultures, different peoples around the world have distinct microbes in their guts. The category of critters known as *Bacteroides* predominates in people who eat high-meat diets (I'm looking at you, United States and Europe), whereas *Prevotella* are more abundant in the guts of those with high-grain diets.[14] But the variation gets more complex than that. For instance, US and European microbiomes are distinct from each other—even people from smaller regions, such as Spain and Denmark, can be distinguished by their microbes[15]—but more similar to each other than they are to microbiomes of people living more traditional lifestyles. Compared to people in the United States, farmers in Malawi, who primarily eat maize, and in Venezuela, who mainly eat cassava, have far more *Prevotella*, consistent with their high-fiber diets but also possibly due to differences in their genetics or environmental exposures.[16] Differences can appear at even smaller scales; for example, Japanese people have genes from marine *Bacteroides* species that degrade seaweed in the gut, perhaps an adaptation to eating sushi.[17] (Of note: These genes were not seen in the St. Louis population that my lab has studied,[18] and all I'll say about that is, if you haven't tried the sushi in St. Louis, I recommend against it.)

You might wonder how, exactly, diet manipulates your microbiome. Well, you'll have to keep wondering, because very few studies have been done at this point on the mechanism—although the connections discovered so far suggest that there might be pervasive effects between diet and malnutrition, infection risk, and acne.

Next, we come to environmental influences on the microbiome, which are profound in childhood, because, well, look: Have you ever watched little kids? They stick their fingers into everything and then put those disgusting fingers into their mouths. It turns out that this isn't entirely a bad thing.

Children who have more diverse microbial communities as youngsters—those exposed to a range of influences such as siblings, pets, or living on or near a farm—tend to have lower rates of immune system defects, such as hay fever, than children who grew up in cities do.[19] Even as adults, we share a lot of microbes with our family members, including our furry family members. Just as we can match you up to the computer mouse you use by the microbes you leave behind, we can also match you to your cohabiting partner, or to your dog, with reasonable accuracy, by the microbes you share.[20]

Most of the things we do don't change our microbes very much because our microbiomes remain distinct even as they age. Your microbiome will be as different from that of your neighbors on your first day of kindergarten as it will be when you both retire. I put together a video mapping out the normal day-to-day variation in two people, who happen to be Amanda and me. We both took sample swabs from our bodies every day for six months. (She's put up with a lot in the name of the microbiome.) She's stopped, but I'm still going more than five years later. The video illustrates that we maintained our separate microbial identities over this six-month period,[21] even though we lived together and had all kinds of exciting and glamorous opportunities to exchange microbes. For each body site, our microbes stayed distinct, although there is a considerable amount of variation day to day. The things we did differently during the six months—traveling to new places, eating exotic meals, and so on—didn't have much of an effect compared to the overall differences between our microbiomes.

Late in life, people tend to have more diverse gut microbial communities overall. (At least that's true among the healthy elderly; low-diversity microbiomes have been associated with both inpatient living in hospitals and with worse health outcomes.[22]) In one respect, however, the last days of our lives resemble our earliest: proteobacteria such as *E. coli* and its relatives tend to be more common both in the elderly and in infants. We don't know yet why this is. It could be because they are recolonizing the sick guts of the elderly and colonizing the underdeveloped guts of infants—proteobacteria tend to be the fast-growing weeds of the microbiome.

If you had to replace your microbiome, would you want to have the microbes from a centenarian, or from a child, or from someone your own age? It's possible that centenarians have especially healthy gut microbiomes, and that's how they've reached old age. On the other hand, it's also possible that their gut microbes, despite doing heroic service, are whipping their last flagella, and, therefore, transplanting them would be inadvisable. Similarly, transplanting a youthful microbiome might seem like a good way to obtain a young and vigorous community apt to develop normally. But suppose a microbe had a beneficial effect at a young age but a detrimental effect at older age. So little research has been done that, at this point, the science doesn't help us. For now, you should probably hold off on any exploratory stool transplants. (More on this in chapter 5.)

3 In Sickness and in Health

As both a scientist and a human being, I am continually awestruck by discoveries about the power of the microbiome to define and shape us. But what excites me most is the very real prospect that, as we come to better understand and even influence the microbiome, it could have the power to heal us.

We're already starting to link our microbes to a wide spectrum of specific diseases, from the obvious—like infectious diseases and inflammatory bowel disease—to surprising ones such as multiple sclerosis, autism, and depression.

It's worth noting that just because we know a microbe is involved in a specific disease, it doesn't mean the answer—or the cure—is to eliminate that microbe. In fact, doing so might cause irreversible damage. It may turn out that targeting diet or inhibiting an enzyme (that's a protein that speeds up a particular chemical reaction) might be more effective than attacking the microbes directly. And yet the reason there is so much excitement about the microbiome is the prospect of discovering entirely new mechanisms to treat conditions that have resisted existing therapies.

But first, let's ask: How is it we know that certain microbes are associated with particular diseases?

The easiest cases to make are those where one particular microbe has a significant impact on health, which essentially describes the last 150 years of infectious disease research. If you get exposed to a microbe such as *Salmonella*, or *Giardia*, or *Mycobacterium tuberculosis* (the bacterium that causes tuberculosis), you expect to get sick. And then, if you treat it with the right antibiotic (or other drug), you expect to get better.

But wait: Do you *always* get sick just because you're exposed?

Actually, our risk of sickness depends on a combination of exposure, genetic makeup, and other factors. Some people are born with resistance to certain diseases. You've probably heard of Typhoid Mary, a New York cook in the early twentieth century who carried the bacteria that causes the disease typhoid fever. She infected family after family with her excellent cooking, which was laced with a dose of her not-so-excellent microbes. But Mary was never sick. She was naturally immune to the fever she carried inside her. Where does such resistance come from? Well, it's these questions that make mouse studies popular with researchers: besides the fact that we can more ethically give a mouse an infection, we can also manipulate the mouse's genome. From these studies, we've learned that susceptibility to essentially every kind of infection hinges greatly on genetics. And mouse versions of Typhoid Mary are easy to create in the lab—not just for typhoid fever but also for a whole range of other infections. It's proof that our genes influence which microbes make each of us sick.

We're beginning to realize that there may be many more diseases where we're all exposed to the same microbe, but it's dangerous only to some of us. We still need more research to explain *why*.

But in the meantime, what follows is a roundup of the key diseases in which we now suspect that microbes may play a part.

Inflammatory Bowel Disease

Inflammatory bowel disease (IBD) is a catchall diagnosis for inflammation of the digestive tract. The big illnesses that fall under the IBD label are ulcerative colitis and Crohn's disease. What these diseases have in common is an altered relationship between intestinal microbes and the immune system. In an attempt to target the pathogens afflicting you, your body goes to war with all the creatures in your intestines, and the intense pain, bleeding, and all-too-frequent trips to the toilet are the collateral damage.

One typical sign of these diseases is an increase in the abundance of certain bacteria. What's particularly interesting is that the microbes in patients do not appear to be behaving normally: their metabolism is off; they're eating and secreting different chemicals. We don't yet know if this altered behavior is caused by the body's immune response or if microbes are at fault. Your immune system does not so much keep lists of good and bad microbes as it concerns itself with good and bad microbe behavior. Your immune system is not the FBI conducting a manhunt for John Dillinger. Instead, it's the guard in the bank who freaks out and opens fire when somebody leaps over the counter and starts stuffing money into a sack.

It's also not clear yet if these inflammatory bowel diseases are caused by a change in the microbiome or if there is something in the genes of the afflicted that causes the body's normal relationship with gut microbes to go awry, and the changes in the microbial population are merely a response. Perhaps it is some combination of both factors?

Celiac disease is related to inflammatory bowel disease and also involves an immune system component: when celiac sufferers eat wheat products, the natural gluten proteins in wheat activate the immune system, which attacks the lining of the gut, shredding it. Celiac was originally identified and named by the Greek physician Aretaeus of Cappadocia in the first or second century AD. But it wasn't as widely known until Dutch physician Willem-Karel Dicke observed in the "Hunger Winter" of 1944–45 during World War II that when wheat was unavailable, his celiac patients survived much better. (Dicke would go on to pioneer the gluten-free diet.) There has been intense interest in whether celiac is linked to the microbiome, but at this point, the dozen or so studies have found essentially no consistent trends associating microbes with celiac. Although many studies are able to find differences between the microbiomes of celiac patients and healthy people, the bacteria in the celiac patients differ from study to study. Clearly the pattern is complex, and more work is needed to understand whether gut bacteria contribute to celiac or simply respond to the altered, gluten-free diets of celiac patients.

Obesity

Until my trip to Peru in 2008, I used to weigh quite a bit more.

Amanda and I hiked the Inca Trail and then spent a week in the Amazon, where we both came down with really nasty diarrhea—not what you want when you're in a tent. We recovered, only to both have it flare up again. To treat it, we both took doses of the same antibiotic. When we got home, we resumed more or less the same diets and exercise patterns we'd had before we left for the trip. However, I lost about eighty pounds in a few months, going from obese to a healthy body weight.

The difference was remarkable. I had to buy new pants, and colleagues took me aside to ask if I had cancer or if there was something else they should know about. In contrast, Amanda lost no weight at all. I believe that the difference was related to a radical change in my microbes: we each responded differently to the same disease and the same course of treatment.

While we can't, of course, draw scientific conclusions from a study of one couple, my experience here mirrors what published studies are increasingly showing. We're learning that there's a strong microbial component to obesity. Normal-sized, germ-free mice that receive a fecal transplant from an obese mouse become fatter themselves. And the experiment works regardless of whether that first mouse was fat because it had been overfed an unhealthy diet[1] or because it had a genetic mutation that made it fat.[2]

You might wonder: Is it the microbes that are doing this or is it something else in the stool? Good question. To answer it, Jeffrey Gordon, a gastroenterologist who directs the Center for Genome Sciences and Systems Biology at the Washington University School of Medicine in St. Louis, and a team of researchers in his lab asked whether you could isolate hundreds of individual strains of bacteria from an individual person, grow each strain in the lab (without the rest of the fecal matter), mix them together in similar proportions as in the original sample, and then transfer the differences in weight by transferring those bacteria to a new host. Indeed they could, proving that it was the microbes responsible for the weight gain—not a virus, an antibody, a chemical, or anything else in the stool. Even more remarkably, by isolating bacteria from lean people, we could design a microbial community that prevented a mouse from gaining the weight it would normally gain when housed with an obese mouse and exposed to its new roommate's chubby microbes.[3]

My lab and others haven't been able yet to design a microbe community that actually slims down a mouse (or a person), although that's certainly the goal. But in yet-unpublished research, other groups have reported using antibiotics to target the bacteria that proliferate on a high-fat diet, successfully slimming down the mice even if they still ate unhealthily.

Many fad diets for humans are now targeted at improving your microbiome. But the evidence that these actually work is limited. We just don't know enough about the ways in which particular microbes affect digestion and absorption to make a targeted intervention. In 2011, researchers at Harvard University published a study in the *New England Journal of Medicine*[4] that found some foods are associated with weight gain, and others with weight loss. It won't shock you to hear that fat-rich French fries are associated with weight gain, more so than any other food. But oddly, the two foods most associated with weight loss are yogurt and nuts, even though both can be high in

SKINNY (BACTERIA-FREE) MOUSE

+

BACTERIA FROM A FAT MOUSE

=

FAT MOUSE

fat. What exactly is going on? Well, microbes might play a role here. We know from studies in mice that particular microbes, or combinations of microbes, are associated with weight gain or weight loss. Could there be a connection between specific foods and the microbes that make us slimmer?

There is plenty of evidence that what you eat alters your microbiome, making it more habitable for some species and less so for others. Gary Wu, a professor of gastroenterology at the University of Pennsylvania, has shown that diet over the long term—a year or more—correlated very strongly with the overall microbiome. It was his team that demonstrated that people who ate a lot of carbohydrates (pasta, potatoes, sugars) tended to have a lot of *Prevotella*. In contrast, people who ate a lot of protein, especially meat (a la the Western diet), tended to have a lot of *Bacteroides*. These two genera of bacteria help us digest and metabolize our food, but they thrive on different foods. We've yet to untangle what influence *Bacteroides* species have on such typically Western diseases as obesity and diabetes, but there are some suggestive correlations. It's exciting to think that we could grow ourselves healthier and leaner microbiomes by altering our diets.

Some dietary changes can rapidly alter our microbes. Peter Turnbaugh, a systems biologist then at Harvard University, and his colleagues got some hardy volunteers to either go vegan or to eat essentially a meat-and-cheese diet. Veganism caused little immediate change to their gut microbes. But the meat-and-cheese diet caused big changes overnight, increasing the kinds of bacteria linked to cardiovascular disease, such as *Bilophila wadsworthia*.[5]

So a sufficiently extreme diet can have bad effects quickly:
an open question is whether there's one that exerts good
effects that fast.

Allergies and Asthma

The idea that reduced microbial diversity leads to asthma
and allergies dates back to the work of David Strachan
at St. George's Hospital Medical School at the University
of London. In the late 1980s, Strachan noticed that later
siblings in larger families tended to have lower rates of
hay fever and related allergies, and he suggested that
catching infections from older siblings (especially classic
childhood diseases) might help train the immune system
to target real invaders, not dust mites.[6,7] This idea, known
as the "hygiene hypothesis," essentially suggests that
keeping ourselves too clean can lead to immunological
problems, as our idle immune systems—unchallenged by
the bacterial and viral pathogens that humans co-evolved
with—get restless.

Since Strachan's time, the focus has shifted away
from common infections such as measles, colds, and flu,
which are now believed to be strictly harmful. Instead,
the modern hygiene hypothesis centers on our squeaky-
clean childhoods, which keep us insulated from diverse
microbes from healthy sources, ranging from soil to leaf
surfaces to domestic or wild animals. To understand how
this works, think of your immune system as a radio: if
you're dialed into a specific station you can hear the music
crystal clear, but if you're between stations then random
signals can cause loud and unpleasant static. In a similar
way, the immune system may find something else to latch

onto if there is no signal. If you're lucky it'll be pollen or peanut butter that spikes through the "static," causing allergies, but if you're unlucky the immune system might latch onto your own cells, causing diabetes, multiple sclerosis, or other autoimmune diseases. Translation for parents: you still shouldn't challenge your kid's immune system by encouraging him or her to eat tainted meat, lick a hospital floor, approach a rabid bat, or otherwise expose themselves to likely harmful microbes—but the modern hygiene hypothesis says encountering good microbes via dirt and contact with healthy, diverse people and animals may be good preventive medicine.

What's the evidence for this? Well, it's been growing rapidly, with more than one in four of the articles on record published just in 2014. Erika von Mutius at the Children's Hospital of the University of Munich is a pioneer in this area. She has showed that exposure to farming in early life reduces the risk of allergies and asthma substantially,[8] and that some of this effect can be explained by children coming in contact with straw, cows, farm milk, and certain bacteria and fungi.[9,10,11] What about the impact of our invariably dusty homes, which seem to harbor all sorts of nasal irritants despite our best efforts with the mop? Contrary to expectations, von Mutius and others have demonstrated that exposure to allergens such as dust mites and cat hair does not explain the incidence of asthma.[12,13]

Some intriguing recent findings suggest that microbial exposure during pregnancy, not just during childhood, may be important for reducing allergic diseases[14] (although some caution is warranted here because in mice,

viral attack or even simulated viral attack during pregnancy can trigger symptoms resembling autism).[15] Other
promising, yet still preliminary, results show that:

- several probiotics can relieve atopic disease and
  asthma[16] (*Lactobacillus salivarius* LS01, in particular,
  can reverse atopic dermatitis symptoms in some
  children).[17]
- changing animals' microbiota with antibiotics can
  induce allergic diseases.[18]
- certain microbe species can reverse food allergies in
  mice[19] or prevent the food allergies from developing
  in the first place[20]—while others can cause them.[21]

The data on whether breast milk can reduce the incidence of these diseases is somewhat equivocal: the few
robust studies that have been done tend to show modest
if any effects.[22,23] Interestingly, simply living in a setting
with more diverse microbes (say, a home with a backyard
garden rather than an urban apartment far from any
parks) seems to decrease risk of allergic disease.[24] And it's
clear that one's setting exists indoors, not just outdoors.
Early exposure to dogs, especially prenatally[25] and in the
first year of life,[26] appears to decrease allergy risks later.
Surprisingly, we showed that having dogs, but not having
kids, increased human microbial diversity for couples
living together.[27] However, exposure to dogs and cats in
adolescence increases the risks of asthma and eczema.[28]
   It's tricky to add up all this early evidence into a
prescription for lowering your child's risk of asthma and
allergies. I'd sum up the recommendations like this: have

a dog (but make sure you start early, ideally prenatally), live on a farm where your kids are exposed to cows and straw, avoid antibiotics early in life, and perhaps take probiotics and breast-feed (although the evidence for those last two is preliminary at present). In general, exposure to diverse microbes, whether through older siblings, pets, or livestock—or through good old-fashioned playing outdoors—seems to help, even if scientists are still sorting out the specific microbes involved. It may be that diversity itself is most important.

Kwashiorkor

Far from the developed world's struggle with its waist-lines, ongoing studies are helping us to better understand the microbial component in a profound cause of human suffering. Kwashiorkor is a disease infamous for the distended bellies that protrude from the famine-racked frames of those it afflicts. For a long time, kwashiorkor was thought to be a form of malnutrition that occurs when a diet lacks protein.

Such malnutrition is prevalent in countries with high levels of food insecurity, which is the technical term for people not having reliable access to nutritious food they can afford. So can't the problem be fixed by just giving people more food? Not always. Providing more calories in the form of rice or corn doesn't work. What does work is using a peanut butter–based supplement fortified with sugar, vitamins, and micronutrients—a supplement that one study shows can rescue 85 percent of malnourished kids who are treated in sub-Saharan Africa. But what about the remaining 15 percent? The peanut butter–based supplement doesn't work for them. The reason appears to be that kwashiorkor is not entirely a disease of malnutrition but also one of the microbiome. The study shows that the peanut-butter supplement can be made more effective for more children when combined with an initial dose of antibiotics to kill off the bad microbes in the sick children.[29]

Even more remarkably, sometimes microbes are *more* important than diet. Many of these studies are carried out in Malawi, where Gordon conducted research, because food insecurity is rampant and the rate of identical twins is very high. Gordon's lab took fecal samples from identical twins on the same diet, where one was healthy and one had kwashiorkor. They then put the microbes from these samples into genetically identical and germ-free mice. The mice that got the microbes from the healthy twin did fine. However, the mice that received the microbes from the twin with kwashiorkor lost 30 percent of their body mass within three weeks and died if not treated. They could be rescued, however, with the same peanut-based supplement that's used with children in the clinic, with a corresponding change in their microbiomes to a healthy one.[30] This suggests strongly that instead of a protein deficiency causing kwashiorkor, as has been long thought, the disease itself resides in the microbiome, waiting to be triggered by food scarcity.

It's ironic that our microbiomes can inflict upon us either stubborn obesity or persistent undernourishment. We can only hope that this knowledge can bring us together to solve the problems of the developed and underdeveloped countries alike.

There are some general trends emerging about links between diseases and the microbiome. For one, we've learned that gut microbial communities that are lower in diversity have been associated with obesity,[31] inflammatory bowel disease,[32] and rheumatoid arthritis.[33] Forgive me if I sound like a platitudinous commencement speaker, but in (microbial) diversity, there is strength. Just as a person who listens to only one very specific type of music or one brand of partisan politics is ill equipped for other types of music or unfamiliar conversations, ill-equipped is the body that has not already coped with a multitude of microbes.

We've also observed that the kinds of bacteria that cause inflammation in the body—especially the proteobacteria (*E. coli* and its relatives) and some kinds of the Firmicute *Clostridium*—are associated with health problems such as persistent diarrhea, inflammatory bowel disease, and, in some studies, obesity. And there are individual pathogens, like *Vibrio cholerae*, which causes cholera, that you might want to worry about too. However, just because you have a particular organism doesn't mean that it will cause a problem for your individual microbial ecosystem.

But all of this raises another interesting question: Does the influence of our microbes extend beyond our guts?

4 The Gut-Brain Axis: How Microbes Affect Your Mood, Your Mind, and More

It's one thing to learn that the microbes in our guts have a say in how sick or well we are, or what our waistlines look like. But our minds, our moods, our behavior—the things that make us ourselves—surely these are inviolably human?

Well, maybe not.

It might sound crazy, but there's increasing evidence that our chorus of microbes gets a say in who we become and how we feel. How might microbes shape our behavior? It turns out that, rather than too few mechanisms, there are almost too many to contemplate.

From their throne in our guts, microbes not only influence how we digest food, absorb drugs, and produce hormones, but they can also interact with our immune systems to affect our brains. Together the various interactions between microbes and the brain are called the microbiome-gut-brain axis,[1,2] and understanding this axis could have profound implications for our understanding of psychiatric disorders and our nervous system.

For example, depression is now known to involve an inflammatory response, and many beneficial bacteria in the gut produce short-chain fatty acids like butyrate, which help feed the cells lining the gut to reduce inflammation. Very recently, the microbiome has been linked to depression in humans, with the discovery that the bacteria *Oscillibacter* produce a chemical that acts as a natural

tranquilizer, mimicking the action of the neurotransmitter GABA, which calms nervous activity in the brain and can lead to depression.[3] The ability of soil microbes such as *Mycobacterium vaccae* to modulate the human immune system has long been known, leading some researchers to suggest that it might be possible to use them to vaccinate against stress and depression.[4] In particular, Graham Rook at University College London has proposed that not having enough contact with our "old friends"—soil microbes that humans had been exposed to throughout human history but that we now isolate ourselves from by clean living—could explain the rapidly increasing frequency of diseases involving inflammation, such as diabetes, arthritis, and even depression.

Moreover, with all of their influence on our body chemistry, microbes may be able to shape our minds as we develop. Autism is an especially interesting case. Several studies have reported that children with autism spectrum disorders differ from neurotypical children (often siblings) in their gut microbiomes.[5] However, because autism is often linked to gut disorders such as diarrhea, which themselves change the microbiome, it can be hard to tell if the differences are due to the autism or to the diarrhea.

Sarkis Mazmanian, a true visionary and a MacArthur Fellow who teaches microbiology at the California Institute of Technology, better known as Caltech, has created a spectacular microbiome-based treatment for autism-like symptoms in mice. Now where, you might ask, does he find a supply of autistic mice? Mazmanian creates them. To do this, he injects pregnant mice with double-stranded RNA, which is chemically similar to DNA but plays different roles in the cell. From the point of view of the mother mice's immune systems, this looks like a virus. Their immune systems then go into overdrive, elevating their body temperature and cytokine levels and, in the crossfire, killing off a lot of their normal microbiota. These mice then give birth to offspring with immune systems and microbiomes that are different from those of normal mice. And it turns out that these offspring have a suite of symptoms that resemble autism in humans. They have cognitive deficits. They have social deficits—they'd rather be on their own than be with other mice. They exhibit repetitive behavior, obsessively burying marbles. And they have gastrointestinal issues.

Mazmanian found that some of these symptoms appear to be due to a molecule called 4-EPS, which is produced in excess by the altered microbiome. Injecting normal mouse pups with 4-EPS re-creates autism-like symptoms. And giving the mice a probiotic strain of *Bacteroides fragilis* reverses some of the symptoms, including the GI issues and the cognitive deficits.[6] Now, before you go looking for *B. fragilis* at the store, keep in mind that the same strain of bacteria that is beneficial to one species can be lethal to another. Until controlled human trials are completed, it's premature to ingest a probiotic for autism— and even unsafe.

That said, the idea that we can isolate the chemicals responsible for a specific condition—even one involving the brain—and then identify the bacteria that either produce or eliminate these chemicals, is very exciting.

Our microscopic passengers can also influence what we do and how we think. And sometimes our genes determine which bacteria live inside us, and then those bacteria turn right around and influence how we behave. This is very well demonstrated in mice lacking a gene called *Tlr5*, which makes them overeat and subsequently become obese. Mice missing *Tlr5* have microbes that make them hungrier; they overeat and become fat. We can prove it's the microbes doing this in two separate experiments. In one, we transfer the *Tlr5*-less mice's microbes into other genetically normal mice, which then overeat and become fat. In the other study, we use antibiotics to wipe out the microbes in the *Tlr5*-less mice and watch as their appetites return to normal. It's amazing to think a genetic tweak can create gut microbes that affect behavior and that this behavior can be transferred into another stomach and alter the behavior of its formerly normal host.[7]

Appetite isn't the only behavior that microbes influence. Anxiety is another. Swapping the microbes between two genetically distinct strains of mice also swaps their performance in anxiety tests. Less anxious mice that get microbes from more anxious mice become more anxious themselves, and, correspondingly, the microbes from less anxious mice can calm more anxious ones.[8] Sven Pettersson, a microbiologist at the Karolinska Institute in Sweden, came up with an elegant way to test this reaction.

Pettersson saw higher anxiety among germ-free mice—those raised in a bubble without any microbes of their own—than among normal mice. But if he transferred the normal bacteria to mice early on, within a few days of birth, they grew up to behave the same way that normal mice did. In contrast, if they were colonized only weeks afterward, they behaved anxiously, like germ-free mice.

Here we see that, at least in mice, microbes act in early childhood to alter behavior irreversibly.[9]

Specific probiotics have also been shown to alter behavior, both in mice and in humans. There are now more than five hundred studies linking probiotics to behavior, especially anxiety and depression. For example, the probiotic *Lactobacillus helveticus* can decrease anxiety in mice,[10] and *Lactobacillus reuteri* can reduce the likelihood that mice will develop infections when they're stressed.[11] *Lactobacillus rhamnosus GG* has been reported to reduce obsessive-compulsive behaviors, such as marble burying, in mice,[12] and as we mentioned in the section on autism, probiotic strains of *Bacteroides fragilis* can rescue mice from some autistic-like traits, including cognitive deficits and repetitive behavior.[13]

Curing mice is all very well, but at some point, you want to be able to make people better too, which biomedical science is pursuing. Clinical trials of certain probiotics have been reported successful: examples include the commercially available probiotics VSL#3 and LCR35 for irritable bowel syndrome [14,15] and *Bifidobacterium infantis natren* for early-life celiac disease.[16] (Irritable bowel syndrome and celiac are both frequently associated with major depression, with some studies reporting that as many as 40 percent of celiac patients are also depressed, further suggesting gut-brain connections.) There has also been at least one report using probiotics to alleviate chronic fatigue syndrome.[17] A cocktail of *Lactobacillus helveticus* and *Bifidobacterium longum* was reported to improve mood in healthy human volunteers.[18] Although this research is still in its early stages, the evidence for psychological effects of altering the microbiome, even in humans, is looking very promising. It's a common personal experience that changing your diet can change your mood. Because changing your diet also changes your microbes, it's entirely possible that some of these effects have a microbial component.

And if microbes can change our health and our minds, the next question is, can we change our microbes to improve ourselves?

A Brief History of Bugs

In the last half of the seventeenth century—for the first and possibly only time in all of history—the Dutch city of Delft was important. There was a new and lucrative trade manufacturing imitation Chinese porcelain called Delftware. The artist Johannes Vermeer was creating vivid works that would, centuries after his death, become some of the most valuable paintings in existence. But most important of all was the hobby of a draper named Antonie van Leeuwenhoek.

The son of a basket maker who had apprenticed himself to a textile merchant, Leeuwenhoek encountered his first magnifying glass, using it to inspect merchandise. What fascinated him most, though, weren't the fabrics but the specialized glass he examined them with. Eventually, Leeuwenhoek taught himself to blow glass and grind magnifying lenses, which let him peer into the small corners of the world. What he found swimming in drops of water were tiny creatures he called animalcules. It was the beginning of microbiology.

It was also, although it took a couple of hundred years for the news to get around, the end of some persuasively bad ideas in medicine. In Leeuwenhoek's time, the understanding of illness revolved around bodily humors and miasmas. Humors were sort of a combination of the periodic table, horoscopes, and a diagnostic chart. The idea was that there were four basic elements—earth, fire, water, and air—that corresponded to four human moods and four substances within the body: black bile, yellow bile, phlegm, and blood. Illness and emotional states were said to be caused by an imbalance of humors. Treatment involved restoring balance, often through bloodletting, induced purges, cupping, or, if you were lucky, a change in diet. At the same time, it was thought that disease was spread by miasma, or polluted air, like that wafting off decaying bodies or out of a swamp. Compared with supernatural theories of disease and divine punishment, humors and miasmas not only sounded reasonable but also offered practical and useful advice, like, say, avoiding the night air from a swamp, which works even if you don't know that mosquitoes carry diseases. (Malaria comes from *mala aria*: medieval Italian for "bad air.") It's just that they happened to be wrong.

With his microscope in the 1670s, Leeuwenhoek discovered a mechanism for disease. One of his first ideas was to see whether his animalcules could explain the differences in oral health between people who cleaned their teeth regularly and those who didn't. He first looked at scrapings from his own teeth and those of two female subjects widely believed to have been his wife and daughter (a subject selection process that university review boards of today would surely frown upon). Then he went out onto the streets of Delft and found two men who swore they had never brushed their teeth in their lives. From the samples he obtained from those men came the first report of bacteria associated with the human body.

By the 1680s, Leeuwenhoek had discovered that there

are different microbes in different parts of the body, and that children have different microbes from adults. And Leeuwenhoek, studying his effluence during a bout of the runs, even showed that microbes were associated with a specific disease, producing descriptions recognizable as *Giardia*, the genus of eukaryotic parasites known to generations of North American nature seekers as "beaver fever."

(Leeuwenhoek was also the first person to microscopically observe sperm. It was a significant accomplishment, although his contemporaries were not terribly enthusiastic about it, despite his assurances that he did not obtain his samples through any abominable acts but rather fresh from his own marital bed.)

The idea that disease was linked to contagious substances that could pass from person to person, however, predates Leeuwenhock. Why didn't his discovery immediately give rise to germ theory? Well, for one thing, at the level of magnification he employed, it was hard to distinguish between different microbes. For another, Leeuwenhoek may have shared his microscopes with others, but he didn't sell them widely and didn't tell anyone how he was making what were at the time the world's finest lenses—a secret he took to his grave.

What held back germ theory was another prevalent and persuasive bad idea: the spontaneous generation of life. It was widely believed that life could arise from nonliving material, with worms popping out of soil and maggots from meat as easily as dew upon a daffodil. Under this theory, microbes, even if they varied in different disease states, might not be important. Perhaps the disease itself created them, as much a symptom as a rash or a pustule.

It wasn't until nearly two hundred years later that several key pieces of the puzzle came together to provide the modern understanding of infectious disease.

In 1847 Ignaz Semmelweis, a Hungarian physician, performed his groundbreaking work, showing that the mortality rate of mothers was reduced dramatically if doctors sterilized their hands in between touching corpses and delivering babies. Semmelweis's medical contemporaries ridiculed his findings: he lost his position in Vienna General Hospital's First Obstetrical Clinic and eventually was committed to an asylum, where he was beaten, and then swiftly (and ironically) contracted a fatal infection.

Seven years after Semmelweis's discovery, English physician John Snow noted that cholera clustered with drinking water and not with bad air, as was previously believed. He traced the origin of a cholera outbreak in London to a single pump on Broad Street. The handle of the pump was subsequently removed—although bureaucracy being what it is, not until after the epidemic had ended. A later committee report dismissed Snow's theories and said it "seems impossible to doubt" that miasmas caused cholera outbreaks.[1]

French chemist and microbiologist Louis Pasteur provided a mechanism to explain all of this. In 1859 Pasteur showed that sterile nutrient broth in a sealed flask couldn't spontaneously generate life; growth occurred only if the flask was broken, exposing the broth to microorganisms in the air—an experiment that gave us the germ theory of disease.

In 1865, after reading Pasteur's work, a British surgeon named Joseph Lister developed antiseptic methods that

increased dramatically the likelihood of his patients'
survival—techniques that, together with antibiotics, essentially make modern surgery possible.

Robert Koch, in 1877, wrote the first rules for linking
a specific microbe to a specific disease. The German
bacteriologist's idea was that in order to prove a microbial
cause, you must be able to find a microbe present in every
person with a disease and not present among the healthy.
Then you must be able to grow the implicated microbe in
pure culture and use a sample from that culture to infect a
healthy host. Then, to be really certain, you have to be able
to take a sample from this newly diseased person and grow
from it a new culture that matches the original microbe.

If you can prove a microbe causes a disease to the
standards of Koch's postulates, then you've *really* proved it.
Doing so isn't easy. As hard as it is to find human volunteers who will let you infect them with a disease, it's impossible to find a university review board that will let you do
so, which is why modern DNA-sequencing technology is so
important. It reveals the great many microbes that reside
inside of us, without sickening volunteers or growing
cultures in the laboratory.

5 Hacking Your Microbiome

Given all that our microbiomes do for us and to us, it's worth asking: Can we build ourselves better ones?

We should be able to. We alter our microbiomes all the time. If you alter the balance of grains and proteins you eat, or change your alcohol intake, you're altering your microbiome. You change it if you use antimicrobial soap or are prescribed a course of antibiotics.

But what if we did it purposefully? What might a microbe-focused medicine look like?

It might help to think of your microbiome like your lawn.[1] Suppose you start with a flourishing lawn, with a bit of diversity—maybe some clover along with the grasses. To keep it growing nicely, with the grass maintaining its dominance over the clover, you might want to fertilize it. Which is where prebiotics come in.

Prebiotics

You may not have heard of them, but prebiotics are like fertilizer for your microbes, providing nutrients they need and that favor the beneficial species. Prebiotics are mostly soluble fibers such as fructans (for example, inulin, lactulose, and the tasty-sounding galacto-oligosaccharides), which naturally occur in some fruits and vegetables. These are fermented by bacteria living in your large intestine, such as *Ruminococcus gnavus*, to produce short-chain fatty acids like butyrate, which provide nutrition for the cells that line your gut.[2] Prebiotics are thought to reproduce some of the benefits of natural high-fiber diets, similar to those our ancestors consumed, by stimulating health-promoting microbes.

Unfortunately, there is no single definition for prebiotics. According to the International Scientific Association for Probiotics and Prebiotics, prebiotics are "nondigestible substances that provide a beneficial physiological effect for the host by selectively stimulating the favorable growth or activity of a limited number of indigenous bacteria."[3] There have been a few randomized, controlled clinical trials[4] (the kind of study that produces the most reliable results) of prebiotics that showed some benefit for Crohn's disease,[5] constipation,[6] and insulin resistance,[7] but most clinical trials to date are still at the stage of proving safety, and the numbers of participants are often too small to draw reliable conclusions about what you should do.

Probiotics

So your lawn is looking lush for a while, but then something terrible happens: a flood wipes it out, or it gets overgrown with crabgrass and dandelions. What do you do? It could be time to reseed it selectively.

Probiotics are mostly bacteria found naturally in the human gut, or in fermented foods such as yogurt. Examples include various species of *Bifidobacterium* and *Lactobacillus*. Probiotics are defined as live microorganisms that, when administered in sufficient quantity, benefit health. Probiotics are referred to as "good bacteria" or "helpful bacteria" and are available as dietary supplements, yogurts, and suppositories. Some probiotic products contain a single strain of bacteria, while others contain a cocktail of different species of bacteria or fungi. The US Food and Drug Administration (FDA) has not yet approved any health claim for probiotic products, so they are marketed as food supplements. (Buyer beware! See below.)

There have been several clinical trials for probiotics, with interest picking up over the past few years, as we get better at reading the microbiome. Some of the most robust evidence exists in support of the preventive and therapeutic effects of probiotics on diarrhea in children[8] and irritable bowel syndrome in adults.[9] Promising applications include prevention and reduction of a severe intestinal condition of premature newborns called necrotizing enterocolitis. Other potential future applications include use in treating obesity, reducing cholesterol levels, and managing irritable bowel syndrome. There are a whole range of possible effects of probiotics, including producing antimicrobial compounds and keeping out harmful bacteria by competing for nutrients and prebiotics. Interestingly, the probiotics don't necessarily have to survive to have an effect—sometimes they alter gut bacteria behavior as they pass through.[10]

One problem with probiotics, as things stand right now, is that there's a lot more hype than there is solid research.

Have you looked at the probiotics section of your supermarket recently? It's possible that the Whole Foods near me in Boulder, Colorado, is an extreme example of this, but what I see there is an entire wall devoted to microbes that are supposed to improve the health of your gut. Unfortunately, the actual evidence that any of these strains will work for anyone is lacking. Although the principles that led to many of them being isolated initially are sound (for instance, they produce short-chain fatty acids such as butyrate), most have not been proved to work for these conditions. It's also unclear whether or not the preparation you can buy at the supermarket contains any live organisms after being shipped there and sitting on the shelf; microorganisms require very specific conditions to survive.

The biggest problem is that many people seem to assume that any probiotic will do. We wouldn't do this for anything else. Suppose you told a friend or relative, "I wasn't feeling well. But I heard that drugs might help. So I took a drug, and I felt better." You might have some follow-up questions. Like: "Which drug did you take?" "Why did you take that particular drug and not some other one?" "Is there any evidence that this particular drug works for the medical condition you have?" Or: "What street corner did you buy this drug on?"

These questions tend not to be asked about probiotics (or, indeed, other microbiome-based therapies). I had almost exactly this conversation with a close family member recently. She said she had unsuccessfully tried two kinds of probiotics to treat her irritable bowel syndrome, which had flared up after a massive dose of antibiotics. I asked how she was picking the probiotics, and she said that one was recommended by a friend and another by her pharmacist. I suggested she try one that had some randomized, controlled trial data suggesting that it worked for IBS.[11] She complained that it was a lot more expensive, but then the next day reported that it had worked amazingly, much better than what she had taken before. Now, nearly a year later, the probiotic essentially has her IBS under control.

Although one example is always just an anecdote, this does reinforce the point that when it comes to anything medical, science can help. So it's worth asking your physician or pharmacist if he or she can recommend probiotics that have randomized, placebo-controlled trials (again, the most robust of studies) backing them. If not, you can survey the latest research in scientific journals yourself (as of this writing, no patient-centered resource exists that compiles this data). Failing that, live yogurt is unlikely to hurt and has helped many people. But the limited clinical trial data available suggest that even different kinds of live yogurt differ substantially in how much they help.[12]

Fecal Transplants

Sometimes, though, you just have to rip out your lawn and lay down fresh sod.

People with severe gastrointestinal illness can literally crap themselves to death. One such disease is *Clostridium difficile*-associated diarrhea. People with *C. diff* have to go to the bathroom dozens of times a day, and the disease is often life threatening. It's also one of the most prevalent hospital-acquired infections in the United States, where each year it sickens 337,000 people and kills 14,000 of them.[13]

Many people take antibiotics for *C. diff*, but this therapy often fails. A complement or possibly an alternative to antibiotic treatment is instead to give the patient microbes from someone who's healthy. One radical and experimental *C. diff* treatment is called fecal transplantation. It's exactly what it sounds like: a healthy volunteer, usually a relative, donates a stool sample, which is then diluted and given to the patient. There are two ways to transplant the feces: the northern route and the southern route. Both routes are effective, curing 90 percent of *C. diff* patients.[14]

Work I did in collaboration with microbiologist Mike Sadowsky and physician Alex Khoruts, both at the University of Minnesota, showed that initially *C. diff* patients have stool communities that look nothing like those of healthy adults, and their fecal microbes resemble those in the skin or vagina. However, within a few days of their fecal transplant, their gut communities are restored to normal, and their symptoms vanish. Fecal transplantation has the power to restore the whole microbial ecosystem in the gut. So far it's been attempted only in dire cases of *C. diff*. But its success has been remarkable, and researchers are very interested in discovering what other conditions it might help. In the lab, as we've mentioned earlier, we've seen that fecal transplants can cure obesity in mice. It'll be exciting to see if we can apply these findings to human therapies.

Vaccines

Continuing the garden metaphor, what if we could keep our lawns from getting sick in the first place?

Vaccination is one of the most effective public health treatments we know of. Vaccines are at least 90 percent effective against diseases they can treat,[15] and they've saved more lives throughout the world than any innovation except clean water.[16]

Vaccines are humanity's greatest triumph in public health. They typically need to be taken only once, or a few times, in childhood, to prevent illnesses from ever occurring in the course of a lifetime. Smallpox has been with mankind since at least the time of the pharaohs,[17] killing untold millions and blinding millions more. Thanks to vaccinations, however, it has been eradicated.[18]

Vaccines are also very specific: they train your immune system to respond only to a particular kind of bacteria—often a species or individual strain—and don't target the rest of your good bacteria. So far vaccines have been used primarily for individual pathogens, starting with the nastiest ones, for obvious reasons. But as the list of vaccines expands, less-lethal kinds of microbes are being targeted, including bacteria and viruses that might kill you decades after they infect you (such as human papillomavirus, or HPV, a known cause of cervical cancer) rather than immediately.

Given what we are now starting to find out about the role of particular kinds of bacteria in various diseases that haven't typically been vaccinated against, could we start to vaccinate against them? For example, could we develop a vaccine against the bacteria that produce a chemical called trimethylamine-N-oxide that leads to cardiovascular problems,[19] or against the *Fusobacterium nucleatum* that is found in tumors in colon cancer,[20] or maybe even against particular kinds of gut bacteria that help us very efficiently—perhaps too efficiently—extract energy from an unhealthy diet and become obese?[21] At this point, these are all simply questions, but the potential is immense.

What about vaccinating against depression or post-traumatic stress disorder (PTSD)? According to the World Health Organization, depression is now the leading cause of disability in the United States and is rapidly becoming more common in the developing world. This increase in depression rates matches the rise of other diseases frequently considered to be Western, such as inflammatory bowel disease, multiple sclerosis, and diabetes, all of which, we now know, have both immune and microbial components. Could our estranged soil bacteria, which modulate the immune system, be playing a role? In experiments in mice, *Mycobacterium vaccae*, a soil bacterium, has reduced anxiety. Intriguingly, in a social stress situation (essentially, smaller mice are put in a cage with a much larger, dominant mouse, which beats them up), *M. vaccae* treatment makes the mice much more resilient against the effects of stress, possibly providing a model for treating stress disorders in humans.[22] Graham Rook, a University College London microbiologist, along with Chuck Raison, a research psychiatrist at the University of Arizona, and Chris Lowry, who teaches integrative physiology at the University of Colorado, proposed possibly creating an *M. vaccae*-based vaccine to treat depression a few years ago[23] and have produced some really promising mouse-model data.

6 Antibiotics

With everything we're learning about the essential, complicated roles that microbes play in almost every part of our lives, we have to ask ourselves: Is it wise to use antibiotics as frequently as we do?

Amanda and I took our daughter to her first doctor's appointment when she was just a few days old. The pediatrician asked us a question with all the care and delicacy of a zoo dentist worried that the lion she's treating might be insufficiently anesthetized. "So," she said, "there are a lot of different opinions about vaccines. How are we feeling about them?"

Amanda and I looked at each other and said, "We would like all the vaccines on the CDC schedule, thank you very much." The US Centers for Disease Control and Prevention publishes a recommended schedule of childhood immunizations.

I don't blame our pediatrician. She was just being sensitive to the (insane) concerns of the community from which she draws her patients. It's exasperating to me how much people worry about vaccines and how little they worry about antibiotics.

Consider what happened when our daughter was born. Right before Amanda underwent an unplanned cesarean, she was given antibiotics. And minutes after our daughter's birth, doctors put antibiotic drops in her

eyes. No one asked—they just did it. This is a standard
treatment designed to guard against the sexually transmitted disease gonorrhea, which can cause conjunctivitis
in infants.[1] We were pretty sure we didn't have gonorrhea.
But the point is, we didn't find out until later about the
antibiotics, which have become so ubiquitous that their
use is not always disclosed before they are administered.
People worry about vaccines, even though almost all
concerns about vaccines are scientifically unfounded or
have been outright disproved. For instance, reports that
certain vaccines cause autism have been comprehensively
disproven, the journal paper that claimed the association
has been retracted, and the author barred from practicing
medicine in his native England.[2] Of course, vaccines do
have risks, but these risks are well documented and minus-
cule: typically one in a million for severe reactions.[3]

Conversely, you hardly ever hear about anyone refusing
antibiotics, despite the fact that they are typically far less
effective than vaccines. While vaccines continue to be at
least 90 percent effective for many diseases, antibiotics
are becoming less effective, in part because of their misuse
and overuse, which is responsible for the rapid spread of
antibiotic resistance,[4] as Marty Blaser, a medical microbiologist at New York University, has outlined so eloquently
in his book on the topic, *Missing Microbes: How the Overuse
of Antibiotics Is Fueling Our Modern Plagues*.[5] (A sobering
fact: more than 70 percent of the bacteria that cause
infections in US hospitals are resistant to at least one of the
antibiotics normally used to treat them.)[6] Blaser argues—
and many agree—that antibiotics are the equivalent of
the chemical weapon napalm. They damage a great many

organisms within us, depleting our microbial heritage in ways we're only beginning to understand, with grave ramifications for health and society.

Antibiotics are essentially poisons that are more toxic to bacteria than they are to us. Bacteria have a lot of biochemical differences from us. Sometimes the difference is in the shapes of molecules that we share with them, like the ribosome, which manufactures proteins. Other times the difference is a molecular machine they make but we don't, such as the enzymes that synthesize their cell walls, which have no equivalent in mammalian cells. Antibiotics target such basic processes—making proteins; dividing; synthesizing the cell wall; transporting nutrients into the cell; and so on—in bacteria. Sometimes antibiotics can punch holes in the cell wall or cell membrane of a bacterium, letting essential components leak out of it like a split sack of groceries.[7]

Antibiotics are relatively safe because they target processes required for microbial life and leave the rest of our cells alone. But there are dangers: in addition to the untargeted destruction of both "good" and "bad" bacteria, we have to worry about bacteria outsmarting the drugs. Pathogens can adapt to antibiotics. Bacterial populations can breed rapidly, allowing them to respond with speed and flexibility to evolutionary pressures. Antibiotics are just such a pressure. Worse, some bacteria have a head start because they've encountered antibiotics before. We don't create antibiotics from scratch; rather, we discover them in the environment. Many of the compounds we use as antibiotics are used first by microbes in the environment, especially soil microbes, to communicate.[8] Because

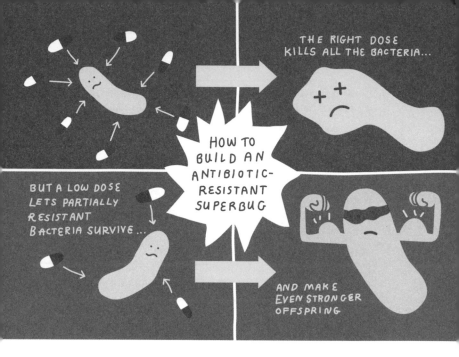

bacteria are already acquainted with these compounds, many microbial species already possess low levels of resistance. But exposure to antibiotics all the time selects for high levels of resistance in all species, including the dangerous ones we're trying to get rid of.

It's not just human-associated bacteria we need to be concerned about. Antibiotic-resistant genes are among those most commonly transferred when bacteria have "sex." Bacteria are incredibly promiscuous, doing it not just across species but also with much more distant relatives (see page 90 on exchanging genetic material outside species). What happens in livestock treated with antibiotics can eventually breed its way back into microbes that live in humans.[9]

It would be one thing if antibiotics were used mostly to

treat sick animals the way they're used to treat sick people. But farmers noticed as far back as the 1950s that livestock on low doses of antibiotics gained weight a lot more rapidly, even at doses lower than the therapeutic dose. Livestock in the United States is commonly treated with low doses of antibiotics solely to increase the size, and thus the value, of the animals.

This is the worst-case scenario for antibiotic resistance. While high doses of antibiotics kill (almost) everything, low doses allow changes that make a bug just a little more resistant, so that when the time comes that a particular bug is indeed life threatening, we've provided it with all the tools and skills it needs to sidestep our attempts to defeat it. Moreover, these bugs survive and spread throughout the agricultural industry and can jump species and infect humans.[10] That's why in 2006 the European Union banned low-dose antibiotic treatment for fattening livestock.

It makes you think: If low doses of antibiotics fatten up our livestock, do they also fatten us up? After all, detectable traces of antibiotics are essentially everywhere in the environment, including our drinking water.

To test this idea, Blaser and his colleagues studied whether mice treated with low doses of antibiotics became heavier than normal mice. Indeed they did, showing that antibiotics affected mice as well as livestock.[11] They also tested whether repeated high doses of antibiotics, like you might use on your kids when they have an ear infection, produced weight gain in mice. Again the answer was yes.[12] In a second branch of the study, Blaser collaborated with epidemiologists—those who study trends in the health of whole populations, not just individuals—to ask whether people who had received antibiotics early in life later put on more pounds than those who didn't. Once again the answer was yes: antibiotics in the first six months were associated especially with increased weight gain.[13] As we saw back in chapter 2, antibiotics can have a profound effect on a child's microbial development, which may account for their apparent influence on later obesity.

I am especially concerned about what antibiotics do to an infant's microbiota. Antibiotic treatment of newborns, even briefly, causes significant alterations to the composition of their gut bacteria. Perhaps more worrisome, antibiotics disturb the normal patterns of colonization of *Bifidobacterium*, one of the beneficial microbes. Colonization by *Bifidobacterium* plays a critical role in the development of a child's immune system. Antibiotic use early in life may thus elevate the risks of allergies and allergic asthma by reducing the beneficial effects of microbial exposure. One large multicenter study found an association between antibiotic use in the first year of life and symptoms of asthma, rhinoconjunctivitis (aka hay fever), and the skin condition eczema in children six to seven years old.[14] And early use of antibiotics could have something to do with the rocketing rates of food allergies among American children. A team based at the University of Chicago showed recently that young mice treated with antibiotics are more likely to develop a peanut-allergy-like condition. But dosing them with certain species from a common type of microbe called *Clostridia* relieved them— apparently by blocking the offending peanut proteins from getting into the bloodstream.[15]

This doesn't mean that you shouldn't take antibiotics,

which save lives and are the only effective treatment option in many circumstances. Ironically, one of the biggest problems with antibiotics is that they often make you feel better almost immediately. This might be why they're so much more accepted by the public than a vaccine. With a vaccine, you take it when you're not sick, and it decreases the risk of illness years later—its effect is delayed and invisible. In contrast, with an antibiotic, you feel sick right now, you take it, and you feel better very quickly. But this in itself is dangerous, because when you start feeling better, there are usually a lot of bacteria still in your system. If you stop taking the antibiotic as soon as you're feeling better, it gives the bacteria that were able to survive the early doses a chance to go on and develop full resistance to that antibiotic. This means that the same antibiotic might not work for you next time, and you could go on to infect other people. So don't try to cut down on antibiotics by not finishing your prescription: if you start, you need to finish the whole course.

We exacerbate the problem because we use unreliable methods to choose the right antibiotic for the job. This was brought home to me when my daughter was about a year old. She'd been having recurring staph infections in her diaper region. It happened again on New Year's Day, so we took her into the clinic, where her pediatrician was out on vacation. The doc who saw her looked at the rash and told us that it might be staph. I said he was probably right, because it was staph the last two times, and here it was again.

On the other hand, he said, it could be strep. In any case, the first-line treatment is the same: take some amoxicillin.

He said they would take a sample and culture it anyway, and in three days we'd know what it really was. We filled the prescription, gave it to her, and the rash cleared up. Antibiotics are amazing when they work.

At eight in the morning on January 3, we got an urgent call from the doctor. He'd come back after the holiday to find lab results showing that our daughter's staph was penicillin resistant. Since amoxicillin is ineffective on such infections, he was alarmed and certain her health must have deteriorated. But the amoxicillin *did* work, and a one-year-old is too young for the placebo effect.

When the doctor learned that my daughter's rash had cleared up, he explained that they test for penicillin resistance but not for *amoxicillin resistance*, and they're slightly different drugs, although they're related. Also, maybe the antibiotic-resistant staph were just on the surface but not in the rash itself. All this underscores how crude our current diagnostic tests are compared with the amazing things we can do in the laboratory. The DNA-sequencing instrument in the building where my lab is located could have completed these tests much faster and with more definitive results. It's not the clinic's fault: the machines and techniques we're using in the laboratory don't yet have FDA approval.

Because bacterial infections are life threatening—and can be hard to diagnose quickly—antibiotics are often prescribed even if the chance of the deadly bacterium they target actually causing the symptoms is low. Add that to high demand from anxious patients (or their parents), and the placebo effect, and they're prescribed far more than they're actually needed. And, to a degree, this makes sense: If we assume that antibiotics are low risk, because they don't have immediate and obvious deleterious effects on our own bodies, why not prescribe them just in case?

Antibiotics can have insidious long-term effects: they become less effective each time you take them and breed antibiotic-resistant bacteria strains that endanger the population as a whole. Plus, broad-spectrum antibiotics such as amoxicillin and ciprofloxacin, which target wide swaths of species, damage our entire microbiome and not just the pathogens we're trying to cure. What would get us out of this mess is better, faster diagnosis. We already have the technology to run tests called polymerase chain reaction (PCR) panels relatively rapidly that can positively identify a range of pathogens. These are especially useful for telling if you have a viral infection, where antibiotics won't help (viruses are not bacteria, so if you've caught the former, then an antiviral medication is the more appropriate prescription). Someday soon, we hope, this technology will make it out of the research laboratory and into hospitals.

If you have a bacterial infection, figuring out whether it's a mild or deadly strain, and if it resists antibiotics, requires laboratory tests using culturing, antibodies, and DNA analysis that can take several days. By then it might be too late. Newer, faster technologies such as mass spectrometry (essentially zapping the sample with a laser and using very accurate scales at the molecular level to weigh the molecules) and better DNA sequencing may accelerate the process and ultimately save lives. These technologies are on the horizon: although they are available in the research laboratory, it will be a few years before they're refined sufficiently for clinical use and win FDA approval. While they may not have been available to my daughter at the start of her life, I'm optimistic that we'll be practicing smarter medicine by the time she's an adult. Because if we can use antibiotics only when they're needed and target infections as narrowly as possible, we can both prolong the usefulness of our antibiotics and do less damage to our microbiomes.

7 The Future

By the time you read this book, we will know substantially more about the human microbiome than we did when the book was written. The rate of progress in microbial science is as astounding as the discoveries we're making: revelations that promise, at every step of the way, to reshape and deepen our understanding of the basic ways our bodies—and even our minds—work.

In just a few short years, we've gone from grasping that our microbial cells outnumber our human cells, to discovering that their genes outnumber ours still further, to understanding that microbes may soon explain all kinds of matters of health and sickness that have heretofore been mysteries. And only in the last couple of years has it become cheap enough for individual people to consider placing their own personal pin on the microbial map, to see how they relate to other people. It's a thrilling time to study microbes—and all the personal swabbing is a small price to pay.

The new microbial frontier extends even beyond our bodies. We're beginning to learn how microbes everywhere are related to one another. The same technologies that allow us to read the human microbiome can also be applied to our pets, livestock, wild animals, and the planet itself. With our newfound knowledge, we can view microbes as a web connecting the health of humans, animals, and the environment—and may eventually

understand how to improve the microscopic ecosystems
we live in and the ecosystems that live in us.

Some of the most exciting potential breakthroughs to
come include:

- tests based on your microbes that tell how you'll
  respond to painkillers, heart drugs, and artificial
  sweeteners;
- a better understanding of how your body, including
  its microbes, responds to diet and exercise, and what
  you personally should to do be healthier; and
- a better understanding of stool transplants: Is every
  one's poop equally curative, or do we need to match
  donors to recipients more precisely? And can we use
  a poop pill instead?

(Okay, perhaps only a microbial scientist could love these prospects.)

And on the longer horizon, we're vigorously pursuing these provocative questions:

- Can we design microbial communities that protect humans from weight gain the way we can for mice?
- Can we design microbes that live on your skin and repel mosquitoes? (A particularly pressing concern for Amanda.)
- Can microbes be used not just to diagnose but also actually to cure the wide range of diseases that we now know they are involved in?

There is still a long way to travel on this voyage of microbial discovery. We are very good at finding out which microbes flourish in a particular ecosystem, but to a large extent, we don't yet know what they're doing, how they're talking to one another, or to us. We also don't know what the unintended consequences of perturbing them are, whether by using antibiotics to wipe out bad bugs or by introducing new kinds of microbes through our diets, our associations with other people and animals, or our contact with the environment. A large part of the challenge right now is that we are changing our microbiomes every day, and we are doing this in an essentially arbitrary and undirected manner. The great power of microbiome science will come when we understand what we need to do in order to achieve a desired effect on the complete ecosystem within us.

To begin building this system, hundreds of scientists

are involved in the Human Microbiome Project, the Earth Microbiome Project, and American Gut, along with literally thousands of members of the general public who have provided both samples (of their feces) and support. The Human Microbiome Project, much like the Human Genome Project, was created to take a genetic census of a healthy microbiome and the deviations from it in a handful of diseases. American Gut aims to broaden that census to include a greater diversity of people in sickness and health. The Earth Microbiome Project hopes to look beyond humans to microbial communities in our planet's ecosystems.

All three projects are groundbreaking, poised to expand our abilities from *description* to *prescription*. The research to come will ultimately yield not just detailed microbial maps of humanity but also a kind of microbial GPS—a guiding body of knowledge that will tell us where we are, as well as where we want to go and how to get there.

The American Gut

What if you want to find out about the microbiome that's specifically *yours*? What if you want to place yourself on the microbial map? Well, do I have an offer for you.

Over Thanksgiving 2012, at a time of year when many of us are thinking about our guts, anthropologist Jeff Leach and I cofounded a project called American Gut.[1] It brings to the general public many of the techniques that my laboratory and others developed for the Human Microbiome Project.

The key breakthrough that made this possible was cheaper DNA sequencing. Now we can offer you, the eager citizen-scientist, a chance to participate in these discoveries at a reasonable price point. Currently, those who donate to the project at $99 or above are entitled to claim as a perk a readout of the kinds of bacteria present in their own microbiome sample, and there are some additional but more expensive perks, available at higher donation levels, which we call "A Week of Feces." The goal of the project is to understand what kinds of microbiomes exist in everyone.

To our knowledge, American Gut is the largest crowd-funded citizen-science project, and as of this writing thousands of people have signed up. All the data are made publicly available to the extent possible while protecting privacy, so that interested researchers, educators, and members of the public can make use of it. And unlike a traditional study, where the data are often locked up for years pending publication of a scientific paper, in American Gut, the data are released immediately, allowing us to find all kinds of new and interesting associations that can be investigated in more carefully controlled studies. The power of the project grows ever greater as more and more people join us in putting themselves on this microbial map.

So send us your dollars (or euros or yen—it's open to international participants), and we'll send you a sampling kit. Then you can send us your feces. Or give the kit as a gift to a loved one or coworker. Finally, a chance to find out what kind of crap they've been full of all along.

The Science (and Art) of Microbiome Mapping

If what you hear about the microbiome can seem quali-
fied or contradictory, it's because this isn't exactly rocket
science—I'd submit it's much harder.

One of our biggest challenges is just figuring out what
we're looking at.

In terms of DNA, humans are essentially identical.[1]
But at the microbial level, our similarities diverge quickly.
The same body part on two people will often harbor very
different microbial species (and even when we share
species, their total populations may vary widely). Choose
two people at random and examine a single microbial cell
from the first person's stool, and then the second person's.
Only about 10 percent of the time will you find a cell of
the same species in both people's stool.[2] In contrast, if you
picked a position on the human genome from those same
two people, their DNA would match 99.9 percent of the
time. Not only is our microbial genome much more diverse
than our human genome, the actual types of microbes
differ vastly from person to person as well.

At this point, you might be wondering how is it that
we know our microbes are so different. A germ is a germ,
right? A *Lactobacillus* is a *Lactobacillus*. Well, not always.
The number of different kinds of microbes you find
depends on how you examine microbes. Simply counting
the various species to determine diversity isn't sufficient.

That's because the number of species we might find can vary depending on how hard we look for them. Think of it like counting fish in a pond. If you go out one afternoon and catch three fish, you can't assume they represent all the fish still in the pond. If you catch two trout and a bass, it doesn't mean that the only fish in the pond are trout and bass and they exist in a two-to-one ratio. What species you find depends on when and how you fish and how many times you visit the pond.

It gets more complicated when we begin to define what actually constitutes a unique microbial species. For animals, it's relatively easy: if two animals can interbreed and produce offspring who are themselves fertile, they are, by definition, from the same species. But microbes don't usually have sex. And when they do, they can exchange genetic material outside their species—so wildly outside it that, for example, bacteria have been shown to exchange with other bacteria, archaea, and even eukaryotes and viruses. Imagine if fish could mate with algae and water skeeters, and it somehow worked! There's another problem: very few microbes will grow in a lab. And yet doing so is required to assign a species an official name and description. It's like reeling in a rare deep-sea fish that explodes before you can land it and identify its species.

There is a work-around. Even though we can't grow most species of microbes in a lab, we can still capture and analyze their DNA. From this we determine that their genomes contain enough diversity to qualify as different species. We can also disregard the concept of species completely and measure diversity using a phylo-genetic tree—like the tree of life sketched by Darwin and later updated by Woese and Fox (see page 5); a microbial community that covers more of the tree is considered more diverse. This is helpful because it allows us to call a pond that includes three species of trout less diverse than one that contains a trout, a bass, and a minnow.

Finally, you have to decide whether you care only about the total number of species, or if you also want to know how many of each species there are (their population) relative to others. Why is this important? Because if you count only the species present, a pond with a trout, a bass, and a thousand minnows would be considered just as diverse as a pond with one of each. Now, depending on what aspect of the ecosystem you're looking at, this might work fine. But it might not. There are many decisions to make when deciding what counts as diversity in a given ecosystem.

Next, we might want to compare ecosystems—the microbial communities that inhabit us—to one another. To accomplish this, microbial scientists use something called unique fraction, or UniFrac,[3] which measures the evolutionary history that separates communities. Cathy Lozupone, one of my first graduate students and now a professor at the Anschutz Medical Campus of the University of Colorado, created the UniFrac technique, delineated elegantly in her doctoral dissertation. With UniFrac, first we plot microbial communities on a phylogenetic tree. Then we use a statistical technique called principal coordinates analysis (PCoA) to calculate the number of possible ways a set of communities can vary from one another on the tree.

If this all sounds like Greek to you—or if it's been a long time since you had any linear algebra (or perhaps never)—not to worry. There are computer algorithms that do the math for us. What's important to remember is that this technique can produce accurate maps to chart the relationships between microbial communities and reveal when similar communities reside nearby on the tree of life.

With this information, we can link the microbiome to specific diseases. The first and most common way we do this is with a cross-sectional study: round up a bunch of sick people and a bunch of healthy people and then compare their microbes. Cross-sectional studies have linked gut microbes in humans to obesity,[4,5] type 1 diabetes,[6,7,8] type 2 diabetes,[9,10,11,12] inflammatory bowel disease,[13,14,15,16] irritable bowel syndrome,[17] colon cancer,[18,19,20,21,22] heart disease,[23,24] rheumatoid arthritis,[25] and a range of other disorders. Cross-sectional studies are very useful because when we spot large differences between healthy and sick populations, we know there is something worth investigating. To determine whether the different microbes actually *caused* the illness, however, we have to design additional studies.

The gold standard for cross-sectional studies is to build a predictive model. Using it, you can take data from some subset of people—sick or healthy—and predict whether the rest of the group has the disease. This has been done successfully for diabetes,[26,27] obesity,[28] and inflammatory bowel diseases.[29] Interestingly, we've seen specific biomarkers (the microbial species or genes involved in a disease) differ in different populations, such as Swedish versus Chinese populations in type 2 diabetes. As a result, it is premature to link individual organisms to individual diseases using what we've learned from cross-sectional studies, because what constitutes a pathogen may vary from population to population.

The Human Microbiome Project is a unique type of cross-sectional study because it's aimed at healthy people rather than people with a disease. The project found that a surprising number—up to 30 percent—of the rigorously screened, healthy volunteers had what we often consider dangerous pathogens, including *Staphylococcus aureus*. The volunteers were, by definition, healthy, suggesting that many people harbor organisms that cause disease only under particular conditions. Think again of weeds: they're a problem only when they grow in the wrong places. This tells us that we may want to switch our focus from "Which are the bad microbes, and how can we avoid contact and/or get rid of them?" to "Why is the same microbe harmless to some people but deadly to others?"

So, once we have some initial clues about microbes that might be involved in certain diseases, we could design a longitudinal study, where you track people over time. Researchers use longitudinal studies when they want to identify subtle effects that produce large changes in the microbiome in some people, while doing nothing to others. To date, remarkably few longitudinal microbiome studies have been conducted. However, we expect to see a lot more of these studies as the cost of DNA sequencing continues to fall.

The gold standard for longitudinal studies is the prospective cohort study. Here volunteers are enrolled while they are healthy or before a treatment is begun (otherwise called an intervention study). You can then ask whether it's possible to predict who gets sick or who will respond to treatment. As of this writing, none of these studies has been large enough to yield definitive conclusions, although TEDDY (The Environmental Determinants of Type I Diabetes in the Young study) is currently generating data for thousands of children at risk for developing diabetes.[30] Prospective cohort studies, especially those conducted over long periods of time, are especially valuable for figuring out if a particular microbiome puts you at risk for developing a disease or for determining whether a treatment would work for you.

There is also the mechanistic study, usually done in mice (for reasons that will soon become obvious), which lets us ascertain how a particular biochemical mechanism works. It generally goes like this: we alter the genes of mice; infuse them with particular chemicals thought to have certain effects; add or remove bacteria; and then examine the effects. Unfortunately, a common requisite for these techniques is that the mice in question must be euthanized and have their internal organs examined.

One of the most useful mechanistic studies is to raise mice in a sterile bubble, where no microbes exist. We can then introduce specific microbes into the mice to see if there is a change. These so-called gnotobiotic mice (from the Greek *gnosis*, "to know," because we know precisely what microbes they've been exposed to) have shown us which microbes affect obesity, or demyelination of neurons (the eroding of nerves' protective sheath) in multiple sclerosis,[31,32] or produce a behavior that resembles autism.[33] It is important to keep in mind, always, that what affects mice might not necessarily affect humans. But the mouse studies still provide invaluable information.

Meanwhile, much of the news you hear about disease in the microbiome can be confusing, contradictory, or sometimes overhyped. What should you believe? When sorting out whether microbes can cure a disease, look for positive results across many studies, especially those that come at the problem from different angles. Such findings are more likely to be true, as are findings from studies with more subjects. In general, what we have now are associations between certain microbes and certain illnesses, backed up by plausible mechanisms (often ones we've worked out in mice) rather than by clear evidence of cause and effect.

Plus, the particular way a study is conducted can have a profound effect on its outcome. For example, if we're looking at irritable bowel syndrome, have we narrowed down our research subjects enough? Are we looking only at IBS patients suffering from gas? From pain? Those who suffer in response to specific diets? Or if we're studying obese patients, we need to ask whether they're insulin resistant. Where is their fat distributed on their body? And so on.

Other factors that need to be taken into account when comparing studies include how samples are stored; how the DNA is extracted from a stool sample; which fragment of the genome is examined for identification; what kind of machine is used for DNA sequencing; what computer software is used for data analysis—and even the settings for that computer software.[34,35,36,37] To see subtle effects, you need very standardized methods.

If this all sounds complicated, it's because the microbiome is a complicated place. And so its study requires thoughtful and deliberate care.

This complexity is worth keeping in mind anytime you hear sweeping claims about it or simple fixes for a variety of its ailments. It's important to ask: Who says so, and how does he or she know? After all, you wouldn't trust a rocket scientist to take you to Mars if he couldn't tell you how far away it was.

ACKNOWLEDGMENTS

I would like to acknowledge the many members of my laboratory, especially Daniel McDonald, Justine Debelius, Jessica Metcalf, Embriette Hyde, Luke Ursell, Amnon Amir, Will Van Treuren, and Dana Willner, and microbiome colleagues including Jairam Vanamala, Marty Blaser, Maria Gloria Dominguez-Bello, Ed Yong, Ruth Ley, Sarkis Mazmanian, Dan Knights, Greg Caporaso, Jack Gilbert, Owen White, Pieter Dorrestein, Nikolaus Correll, Ajay Kshatriya, Andrea Edwards, and Dawn Field. My parents, Allison and John Knight, and partner, Amanda Birmingham, also provided valuable input on the whole book and especially the events where they were present, and I would especially like to thank Amanda and my daughter, Alice, for their patience with this project and for their support. My TED speaker coaches Michael Weitz and Abigail Tenenbaum both improved my TED Talk dramatically and provided a very helpful look over the book from a nonbiologist perspective, and Chris Anderson, June Cohen, and the rest of the TED team really helped me rethink how to connect with the public. Michelle Quint, Michael Behar, and Grace Rubenstein were very helpful getting the book across the finish line. My eight hundred or so collaborators (according to my National Science Foundation conflicts-of-interest list), especially Jeffrey I. Gordon, students and lab members, and colleagues in

BioFrontiers at Boulder, provided inspiration and were very forgiving in putting up with other projects being delayed while the book was written. The work described in this book involves a vast and growing research community, and the research from my lab was funded by the Howard Hughes Medical Institute, the NIH (including the Human Microbiome Project), NSF, US Department of Energy (DOE), Defense Advanced Research Projects Agency (DARPA), National Aeronautics and Space Administration (NASA), National Institute of Justice (NIJ), the US-Israel Binational Science Foundation, the W. M. Keck Foundation, the Alfred P. Sloan Foundation, the John Templeton Foundation, the Jane and Charlie Butcher Foundation, the Colorado Center for Biofuels and Biorefining, the Crohn's & Colitis Foundation of America, the Bill & Melinda Gates Foundation, the Gordon and Betty Moore Foundation, and by thousands of members of the public. All errors and omissions are, of course, my own, although much of the engaging writing comes from Brendan.

NOTES

INTRODUCTION

1. Note that the recent American Academy of Microbiology report narrows this lead to 3:1, mostly by increasing the count of human cells. But either way, our microbes outnumber us substantially. See http://academy.asm.org/index.php/faq-series/5122-humanmicrobiome.

2. Available online from Project Gutenberg, www.gutenberg.org/files/1228/1228-h/1228-h.htm.

3. C. R. Woese and G. E. Fox, "Phylogenetic Structure of the Prokaryotic Domain: The Primary Kingdoms," *Proceedings of the National Academy of Sciences of the United States of America* 74, no. 11 (November 1, 1977): 5088-90.

CHAPTER 1:
THE BODY MICROBIAL

1. N. O. Verhulst et al., "Composition of Human Skin Microbiota Affects Attractiveness to Malaria Mosquitoes," *PloS One 6*, no. 12 (2011): e28991.

2. E. A. Grice et al., "Topographical and Temporal Diversity of the Human Skin Microbiome," *Science* 324, no. 5931 (May 29, 2009): 1190-92; E. K. Costello et al., "Bacterial Community Variation in Human Body Habitats Across Space and Time," *Science* 326, no. 5960 (December 18, 2009): 1694-97.

3. F. R. Blattner et al., "The Complete Genome Sequence of *Escherichia Coli* K-12," *Science* 277, no. 5331 (September 5, 1997): 1453-62.

4. R. H. MacArthur and E. O. Wilson, *The Theory of Island Biogeography* (Princeton, NJ: Princeton University Press, 2001).

5. N. Fierer et al., "Forensic Identification Using Skin Bacterial Communities," *Proceedings of the National Academy of Sciences of the United States of America* 107, no. 14 (April 6, 2010): 6477-81.

6. "*CSI: Miami* Season 9," Wikipedia, http://en.wikipedia.org/wiki/List_of_CSI:_Miami_episodes#Season_9:_2010.E2.80.932011.

7. For a highly informative and entertaining introduction to body farms, see Mary Roach, *Stiff: The Curious Lives of Human Cadavers* (New York: W. W. Norton, 2004).

8. Meagan B. Gallagher, Sonia Sandhu, and Robert Kimsey, "Variation in Developmental Time for Geographically Distinct Populations of the Common Green Bottle Fly, *Lucilia sericata* (Meigen)," *Journal of Forensic Sciences* 55, no. 2 (March 2010): 438–42.

9. O. S. Von Ehrenstein et al., "Reduced Risk of Hay Fever and Asthma Among Children of Farmers," *Clinical and Experimental Allergy: Journal of the British Society for Allergy and Clinical Immunology* 30, no. 2 (February 2000): 187–93; E. von Mutius and D. Vercelli, "Farm Living: Effects on Childhood Asthma and Allergy," *Nature Reviews Immunology* 10, no. 12 (December 2010): 861–68.

10. E. S. Charlson et al., "Assessing Bacterial Populations in the Lung by Replicate Analysis of Samples from the Upper and Lower Respiratory Tracts," *PloS One* 7, no. 9 (2012): e42786; E. S. Charlson et al., "Topographical Continuity of Bacterial Populations in the Healthy Human Respiratory Tract," *American Journal of Respiratory and Critical Care Medicine* 184, no. 8 (October 15, 2011): 957–63.

11. J. K. Harris et al., "Molecular Identification of Bacteria in Bronchoalveolar Lavage Fluid from Children with Cystic Fibrosis," *Proceedings of the National Academy of Sciences of the United States of America* 104, no. 51 (December 18, 2007): 20529–33.

12. E. S. Charlson et al., "Topographical Continuity of Bacterial Populations in the Healthy Human Respiratory Tract," *American Journal of Respiratory and Critical Care Medicine* 184, no. 8 (October 15, 2011): 957–63.

13. A. Morris et al., "Comparison of the Respiratory Microbiome in Healthy Nonsmokers and Smokers," *American Journal of Respiratory and Critical Care Medicine* 187, no. 10 (May 15, 2013): 1067–75.

14. O. E. Cornejo et al., "Evolutionary and Population Genomics of the Cavity Causing Bacteria *Streptococcus Mutans*," *Molecular Biology and Evolution* 30, no. 4 (April 2013): 881–93.

15. J. Slots, "The Predominant Cultivable Microflora of Advanced Periodontitis," *Scandinavian Journal of Dental Research* 85, no. 2 (January/February 1977): 114–21.

16. M. Castellarin et al., "*Fusobacterium Nucleatum* Infection Is Prevalent in Human Colorectal Carcinoma," *Genome Research* 22, no. 2 (February 2012): 299–306; M. R. Rubinstein et al., "*Fusobacterium Nucleatum* Promotes Colorectal Carcinogenesis by Modulating E-Cadherin/Beta-Catenin Signaling via Its FadA Adhesin," *Cell Host & Microbe* 14, no. 2 (August 14, 2013): 195–206; A. D. Kostic et al., "*Fusobacterium Nucleatum* Potentiates Intestinal Tumorigenesis and Modulates

the Tumor-Immune Microenvironment," *Cell Host & Microbe* 14 (2013): 207–15; R. L. Warren et al., "Co-occurrence of Anaerobic Bacteria in Colorectal Carcinomas," *Microbiome* 1, no. 1 (May 15, 2013): 16; L. Flanagan et al., "Fusobacterium Nucleatum Associates with Stages of Colorectal Neoplasia Development, Colorectal Cancer and Disease Outcome," *European Journal of Clinical Microbiology & Infectious Diseases: Official Publication of the European Society of Clinical Microbiology* 33 no. 8 (August 2014): 1381–90.

17. D. Falush et al., "Traces of Human Migrations in *Helicobacter Pylori* Populations," *Science* 299, no. 5612 (March 7, 2003): 1582–85.

18. P. B. Eckburg et al., "Diversity of the Human Intestinal Microbial Flora," *Science* 308, no. 5728 (June 10, 2005): 1635–38.

19. M. Hamady and R. Knight, "Microbial Community Profiling for Human Microbiome Projects: Tools, Techniques, and Challenges," *Genome Research* 19, no. 7 (July 2009): 1141–52.

20. Human Microbiome Project Consortium, "Structure, Function and Diversity of the Healthy Human Microbiome," *Nature* 486, no. 7402 (June 13, 2012): 207–14.

21. Eckburg et al., "Diversity of the Human Intestinal Microbial Flora," 1635–38.

22. R. E. Ley et al., "Microbial Ecology: Human Gut Microbes Associated with Obesity," *Nature* 444, no. 7122 (December 21, 2006): 1022–23; P. J. Turnbaugh et al., "A Core Gut Microbiome in Obese and Lean Twins," *Nature* 457, no.7228 (January 22, 2009): 480–84; J. Henao-Mejia et al., "Inflammasome-Mediated Dysbiosis Regulates Progression of NAFLD and Obesity," *Nature* 482, no. 7384 (February 1, 2012): 179–85; V. K. Ridaura et al., "Gut Microbiota from Twins Discordant for Obesity Modulate Metabolism in Mice," *Science* 341, no. 6150 (September 6, 2013): 1241214; M. L. Zupancic et al., "Analysis of the Gut Microbiota in the Old Order Amish and Its Relation to the Metabolic Syndrome," *PloS One* 7, no. 8 (2012): e43052; D. Knights et al., "Human-Associated Microbial Signatures: Examining Their Predictive Value," *Cell Host & Microbe* 10, no. 4 (October 20, 2011): 292–96; E. Le Chatelier et al., "Richness of Human Gut Microbiome Correlates with Metabolic Markers," *Nature* 500, no. 7464 (August 29, 2013): 541–46; A. Cotillard et al., "Dietary Intervention Impact on Gut Microbial Gene Richness," *Nature* 500, no. 7464 (August 29, 2013): 585–88.

23. R. A. Koeth et al., "Intestinal Microbiota Metabolism of L-Carnitine, a Nutrient in Red Meat, Promotes Atherosclerosis," *Nature Medicine* 19, no. 5 (May 2013): 576–85; W. H. Tang et al.,

"Intestinal Microbial Metabolism of Phosphatidylcholine and Cardiovascular Risk," *New England Journal of Medicine* 368, no. 17 (April 25, 2013): 1575–84.

24. Y. K. Lee et al., "Proinflammatory T-cell Responses to Gut Microbiota Promote Experimental Autoimmune Encephalomyelitis," supplement 1, *Proceedings of the National Academy of Sciences of the United States of America* 108 (March 15, 2011): 4615–22; K. Berer et al., "Commensal Microbiota and Myelin Autoantigen Cooperate to Trigger Autoimmune Demyelination," *Nature* 479 (2011): 538–41.

25. E. Y. Hsiao et al., "Microbiota Modulate Behavioral and Physiological Abnormalities Associated with Neurodevelopmental Disorders," *Cell* 155, no. 7 (December 19, 2013): 1451–63.

26. P. Gajer et al., "Temporal Dynamics of the Human Vaginal Microbiota," *Science Translational Medicine* 4, no. 132 (May 2, 2012): 132ra52; J. Ravel et al., "Daily Temporal Dynamics of Vaginal Microbiota Before, During and After Episodes of Bacterial Vaginosis," *Microbiome* 1, no. 1 (December 2, 2013): 29.

CHAPTER 2:
HOW WE GET OUR MICROBIOME

1. R. Romero et al., "The Composition and Stability of the Vaginal Microbiota of Normal Pregnant Women Is Different from That of Non-Pregnant Women," *Microbiome* 2, no. 1 (Febuary3, 2014): 4.

2. O. Koren et al., "Host Remodeling of the Gut Microbiome and Metabolic Changes During Pregnancy," *Cell* 150, no. 3 (August 3, 2012): 470–80.

3. K. Aagaard et al., "The Placenta Harbors a Unique Microbiome," *Science Translational Medicine* 6, no. 237 (May 21, 2014): 237ra65.

4. Romero et al., "The Composition and Stability of Vaginal Microbiota of Normal Pregnant Women Different from That of Non-Pregnant Women."

5. Michelle K. Osterman and Joyce A. Martin, "Changes in Cesarean Delivery Rates by Gestational Age: United States, 1996–2011," NCHS Data Brief, no. 124, June 2013: 1–8; Luz Gibbons et al., *The Global Numbers and Costs of Additionally Needed and Unnecessary Cesarean Sections Performed per Year: Overuse as a Barrier to Universal Coverage* (Geneva, Switzerland: World Health Organization, 2010).

6. M. G. Dominguez-Bello et al., "Delivery Mode Shapes the Acquisition and Structure of the Initial Microbiota Across Multiple Body Habitats in Newborns," *Proceedings of the National Academy of Sciences of the United States of America* 107, no. 26 (June 29, 2010): 11971–75.

7. G. V. Guibas et al., "Conception via In Vitro Fertilization and Delivery by Caesarean Section

Are Associated with Paediatric Asthma Incidence," *Clinical and Experimental Allergy: Journal of the British Society for Allergy and Clinical Immunology* 43, no. 9 (September 2013): 1058–66; L. Braback, A. Lowe, and A. Hjern, "Elective Cesarean Section and Childhood Asthma," *American Journal of Obstetrics and Gynecology* 209, no. 5 (November 2013): 496; C. Roduit et al., "Asthma at 8 Years of Age in Children Born by Caesarean Section," *Thorax* 64, no. 2 (February 2009): 107–13; M. C. Tollanes et al., "Cesarean Section and Risk of Severe Childhood Asthma: A Population-Based Cohort Study," *Journal of Pediatrics* 153, no. 1 (July 2008): 112–16; B. Xu et al., "Caesarean Section and Risk of Asthma and Allergy in Adulthood," *Journal of Allergy and Clinical Immunology* 107, no. 4 (April 2001): 732–33.

8. M. Z. Goldani et al., "Cesarean Section and Increased Body Mass Index in School Children: Two Cohort Studies from Distinct Socioeconomic Background Areas in Brazil," *Nutrition Journal* 12, no. 1 (July 25, 2013): 104; A. A. Mamun et al., "Cesarean Delivery and the Long-term Risk of Offspring Obesity," *Obstetrics and Gynecology* 122, no. 6 (December 2013): 1176–83; D. N. Mesquita et al., "Cesarean Section Is Associated with Increased Peripheral and Central Adiposity in Young Adulthood: Cohort Study," *PloS One* 8, no. 6

(June 27, 2013): e66827; K. Flemming et al., "The Association Between Caesarean Section and Childhood Obesity Revisited: A Cohort Study," *Archives of Disease in Childhood* 98, no. 7 (July 2013): 526–32; E. Svensson et al., "Caesarean Section and Body Mass Index Among Danish Men," *Obesity* 21, no. 3 (March 2013): 429–33; H. T. Li, Y. B. Zhou, and J. M. Liu, "The Impact of Cesarean Section on Offspring Overweight and Obesity: A Systematic Review and Meta-Analysis," *International Journal of Obesity* 37, no. 7 (July 2013): 893–99; H. A. Goldani et al., "Cesarean Delivery Is Associated with an Increased Risk of Obesity in Adulthood in a Brazilian Birth Cohort Study," *American Journal of Clinical Nutrition* 93, no. 6 (June 2011): 1344–47; L. Zhou et al., "Risk Factors of Obesity in Preschool Children in an Urban Area in China," *European Journal of Pediatrics* 170, no. 11 (November 2011): 1401–6.

9. T. Marrs et al., "Is There an Association Between Microbial Exposure and Food Allergy? A Systematic Review," *Pediatric Allergy and Immunology: Official Publication of the European Society of Pediatric Allergy and Immunology* 24, no. 4 (June 2013): 311–20 e8.

10. J. Penders et al., "Establishment of the Intestinal Microbiota and Its Role for Atopic Dermatitis in Early Childhood," *Journal of Allergy and Clinical Immunology* 132, no. 3 (September 2013):

601–7 e8; K. Pyrhonen et al., "Caesarean Section and Allergic Manifestations: Insufficient Evidence of Association Found in Population-Based Study of Children Aged 1 to 4 Years," *Acta Paediatrica* 102, no. 10 (October 2013): 982–89; F. A. van Nimwegen et al., "Mode and Place of Delivery, Gastrointestinal Microbiota, and Their Influence on Asthma and Atopy," *Journal of Allergy and Clinical Immunology* 128, no. 5 (November 2011): 948–55 e1–3; P. Bager, J. Wohlfahrt, and T. Westergaard, "Caesarean Delivery and Risk of Atopy and Allergic Disease: Meta-Analyses," *Clinical and Experimental Allergy: Journal of the British Society for Allergy and Clinical Immunology* 38, no. 4 (April 2008): 634–42; K. Negele et al., "Mode of Delivery and Development of Atopic Disease During the First 2 Years of Life," *Pediatric Allergy and Immunology: Official Publication of the European Society of Pediatric Allergy and Immunology* 15, no. 1 (February 2004): 48–54.

11. M. B. Azad et al., "Gut Microbiota of Healthy Canadian Infants: Profiles by Mode of Delivery and Infant Diet at 4 Months," *CMAJ: Canadian Medical Association Journal* 185, no. 5 (March 19, 2013): 385–94.

12. J. E. Koenig et al., "Succession of Microbial Consortia in the Developing Infant Gut Microbiome," supplement 1, *Proceedings of the National Academy of Sciences of the United States of America* 108 (March 15, 2011): 4578–85.

13. G. D. Wu et al., "Linking Long-term Dietary Patterns with Gut Microbial Enterotypes," *Science* 334, no. 6052 (October 7, 2011): 105–8.

14. Ibid.

15. J. Qin et al., "A Human Gut Microbial Gene Catalogue Established by Metagenomic Sequencing," *Nature* 464, no. 7285 (March 4, 2010): 59–65.

16. T. Yatsunenko et al., "Human Gut Microbiome Viewed Across Age and Geography," *Nature* 486, no. 7402 (May 9, 2012): 222–7.

17. J. H. Hehemann et al., "Transfer of Carbohydrate-Active Enzymes from Marine Bacteria to Japanese Gut Microbiota," *Nature* 464, no. 7290 (April 8, 2010): 908–12.

18. P. J. Turnbaugh et al., "A Core Gut Microbiome in Obese and Lean Twins," *Nature* 457, no. 7228 (Jan. 22, 2009): 480–4.

19. J. Genuneit et al., "The Combined Effects of Family Size and Farm Exposure on Childhood Hay Fever and Atopy," *Pediatric Allergy and Immunology: Official Publication of the European Society of Pediatric Allergy and Immunology* 24, no. 3 (May 2013): 293–98.

20. S. J. Song et al., "Cohabiting Family Members Share Microbiota with One Another and with Their Dogs," *eLife* 2 (April 16, 2013): e00458.

21. J. G. Caporaso et al., "Moving Pictures of the Human Microbiome," *Genome Biology* 12, no. 5 (2011): R50.

22. M. J. Claesson et al., "Gut Microbiota Composition Correlates with Diet and Health in the Elderly," *Nature* 488, no. 7410 (August 9, 2012): 178–84.

CHAPTER 3:
IN SICKNESS AND IN HEALTH

1. P. J. Turnbaugh et al., "Diet-Induced Obesity Is Linked to Marked but Reversible Alterations in the Mouse Distal Gut Microbiome," *Cell Host & Microbe* 3, no. 4 (April 17, 2008): 213–23.

2. M. Vijay-Kumar et al., "Metabolic Syndrome and Altered Gut Microbiota in Mice Lacking Toll-like Receptor 5," *Science* 328, no. 5975 (April 9, 2010): 228–31.

3. Ridaura et al., "Gut Microbiota from Twins Discordant for Obesity Modulate Metabolism in Mice.

4. D. Mozaffarian et al., "Changes in Diet and Lifestyle and Long-term Weight Gain in Women and Men," *New England Journal of Medicine* 364, no. 25 (June 23, 2011): 2392–404.

5. L. A. David et al., "Diet Rapidly and Reproducibly Alters the Human Gut Microbiome," *Nature* 505, no. 7484 (January 23, 2014): 559–63.

6. D. P. Strachan, "Hay Fever, Hygiene, and Household Size," *British Medical Journal*, 299, no. 6710 (Nov. 18, 1989): 1259–60.

7. D. P. Strachan, "Is Allergic Disease Programmed in Early Life?," *Clinical & Experimental Allergy*, 24, no. 7 (July 1994): 603–5.

8. J. Riedler et al., "Exposure to Farming in Early Life and Development of Asthma and Allergy: A Cross-Sectional Study," *The Lancet*, 358, no. 9288 (Oct. 6, 2001): 1129–33.

9. S. Illi et al., "Protection from Childhood Asthma and Allergy in Alpine Farm Environments—the GABRIEL Advanced Studies," *Journal of Allergy and Clinical Immunology*, 129, no. 6 (June 2012): 1470–7.

10. C. Braun-Fahrländer et al., "Environmental Exposure to Endotoxin and Its Relation to Asthma in School-Age Children," *New England Journal of Medicine* 347, no. 12 (Sept. 19, 2002): 869–77.

11. J. Douwes et al., "Does Early Indoor Microbial Exposure Reduce the Risk of Asthma? The Prevention and Incidence of Asthma and Mite Allergy Birth Cohort Study," *Journal of Allergy and Clinical Immunology* 117, no. 5 (May 2006): 1067–73.

12. S. Lau et al., "Early Exposure to House-Dust Mite and Cat Allergens and Development of Childhood Asthma: A Cohort Study. Multicenter Allergy Study Group," *The Lancet* 356, no. 9239 (Oct. 21, 2000): 1392–7.

13. M. J. Ege et al., "Exposure to Environmental Microorganisms and Childhood Asthma," *New England Journal of Medicine* 364, no. 8 (Feb. 24, 2011): 701–9.

14. T. R. Abrahamsson el al, "Gut Microbiota and Allergy: The Importance of the Pregnancy Period," *Pediatric Research* (Oct. 13, 2014): epub ahead of print.

15. E. Y. Hsiao et al., "Microbiota Modulate Behavioral and Physiological Abnormalities Associated With Neurodevelopmental Disorders," *Cell* 155, no. 7 (Dec. 19, 2013): 1451–63.

16. N. Elazab et al., "Probiotic Administration in Early Life, Atopy, and Asthma: A Meta-Analysis of Clinical Trials," *Pediatrics* 132, no. 3 (Sept. 2013): 666–76.

17. A. A. Niccoli et al., "Preliminary Results on Clinical Effects of Probiotic Lactobacillus Salivarius LS01 in Children Affected by Atopic Dermatitis," *Journal of Clinical Gastroenterology* 48, supplement 1 (Nov–Dec 2014): S34–6.

18. M. C. Arrieta and B. Finlay, "The Intestinal Microbiota and Allergic Asthma," *Journal of Infection* 4453, no. 14 (Sept. 25, 2014): 227–8.

19. A. Du Toit, "Microbiome: Clostridia Spp. Combat Food Allergy in Mice," *National Review of Microbiology* 12, no. 10 (Sept. 16, 2014): 657.

20. A. T. Stefka et al., "Commensal Bacteria Protect Against Food Allergen Sensitization," *Proceedings of the National Academy of Sciences* 111, no. 36 (Sept. 9, 2014): 13145–50.

21. M. Noval Rivas et al., "A Microbiota Signature Associated with Experimental Food Allergy Promotes Allergic Sensitization and Anaphylaxis," *Journal of Allergy and Clinical Immunology* 131, no. 1 (Jan. 2013): 201–12.

22. M. S. Kramer et al., "Promotion of Breastfeeding Intervention Trial (PROBIT): a Cluster-Randomized Trial in the Republic of Belarus. Design, Follow-Up, and Data Validation," *Advances in Experimental Medicine and Biology* 478 (2000): 327–45.

23. H. Kronborg et al., "Effect of Early Postnatal Breastfeeding Support: A Cluster-Randomized Community Based Trial," *Acta Pediatrica* 96, no. 7 (July 2007): 1064–70.

24. I. Hanski et al., "Environmental Biodiversity, Human Microbiota, and Allergy Are Interrelated," *Proceedings of the National Academy of Sciences* 109, no. 21 (May 22, 2012): 8334–9.

25. C. G. Carson, "Risk Factors for Developing Atopic Dermatitis," *Danish Medical Journal* 60, no. 7 (July 2013): B4687.

26. E. von Mutius et al., "The PASTURE Project: E.U. Support for the Improvement of Knowledge About Risk Factors and

Preventive Factors for Atopy in Europe," *Allergy* 61, no. 4 (April 2006): 407–13.

27. S. J. Song et al., "Cohabiting Family Members Share Microbiota with One Another and With Their Dogs," *eLife* 2 (April 16, 2013).

28. B. Brunekreef et al., "Exposure to Cats and Dogs, and Symptoms of Asthma, Rhinoconjunctivitis, and Eczema," *Epidemiology* 23, no. 5 (Sept. 2012): 742–50.

29. I. Trehan et al., "Antibiotics as Part of the Management of Severe Acute Malnutrition," *New England Journal of Medicine* 368, no. 5 (January 31, 2013): 425–35.

30. M. I. Smith et al., "Gut Microbiomes of Malawian Twin Pairs Discordant for Kwashiorkor," *Science* 339, no. 6119 (February 1, 2013): 548–54.

31. Turnbaugh et al., "A Core Gut Microbiome in Obese and Lean Twins."

32. D. N. Frank et al., "Molecular-Phylogenetic Characterization of Microbial Community Imbalances in Human Inflammatory Bowel Diseases," *Proceedings of the National Academy of Sciences of the United States of America* 104, no. 34 (August 21, 2007): 13780–85; M. Tong et al., "A Modular Organization of the Human Intestinal Mucosal Microbiota and Its Association with Inflammatory Bowel Disease," *PloS One* 8, no. 11 (November 19, 2013): e80702.

33. J. U. Scher et al., "Expansion of Intestinal *Prevotella Copri* Correlates with Enhanced Susceptibility to Arthritis," *eLife* 2 (November 5, 2013): e01202.

CHAPTER 4:
THE GUT-BRAIN AXIS: HOW MICROBES AFFECT YOUR MOOD, YOUR MIND, AND MORE

1. P. Bercik, "The Microbiota-Gut-Brain Axis: Learning from Intestinal Bacteria?," *Gut* 60, no. 3 (March 2011): 288–89.

2. J. F. Cryan and S. M. O'Mahony, "The Microbiome-Gut-Brain Axis: From Bowel to Behavior," *Neurogastroenterology and Motility: The Official Journal of the European Gastrointestinal Motility Society* 23, no. 3 (March 2011): 187–92.

3. A. Naseribafrouei et al., "Correlation Between the Human Fecal Microbiota and Depression," *Neurogastroenterology and Motility: The Official Journal of the European Gastrointestinal Motility Society* 26, no. 8 (August 2014): 1155–62.

4. G. A. Rook, C. L. Raison, and C. A. Lowry, "Microbiota, Immunoregulatory Old Friends and Psychiatric Disorders," *Advances in Experimental Medicine and Biology* 817 (2014): 319–56.

5. D. W. Kang et al., "Reduced Incidence of *Prevotella* and Other Fermenters in Intestinal Microflora of Autistic Children," *PLoS One* 8, no. 7 (2013): e68322.

6. Hsiao et al., "Microbiota Modulate Behavioral and Physiological Abnormalities Associated with Neurodevelopmental Disorders."

7. Vijay-Kumar et al., "Metabolic Syndrome and Altered Gut Microbiota in Mice Lacking Toll-like Receptor 5."

8. P. Bercik et al., "The Intestinal Microbiota Affect Central Levels of Brain-Derived Neurotropic Factor and Behavior in Mice," *Gastroenterology* 141, no. 2 (August 2011): 599–609, 609 e1-3.

9. R. Diaz Heijtz et al., "Normal Gut Microbiota Modulates Brain Development and Behavior," *Proceedings of the National Academy of Sciences of the United States of America* 108, no. 7 (February 15, 2011): 3047-52.

10. C. L. Ohland et al., "Effects of *Lactobacillus Helveticus* on Murine Behavior Are Dependent on Diet and Genotype and Correlate with Alterations in the Gut Microbiome," *Psychoneuroendocrinology* 38 (2013): 1738-47.

11. A. R. Mackos et al., "Probiotic *Lactobacillus Reuteri* Attenuates the Stressor-Enhanced Sensitivity of *Citrobacter Rodentium* Infection," *Infection and Immunity* 81, no. 9 (September 2013): 3253-63.

12. P. A. Kantak, D. N. Bobrow, and J. G. Nyby, "Obsessive-Compulsive-like Behaviors in House Mice Are Attenuated by a Probiotic (Lactobacillus Rhamnosus GG)," *Behavioural Pharmacology* 25, no. 1 (February 2014): 71-79.

13. Hsiao et al., "Microbiota Modulate Behavioral and Physiological Abnormalities Associated with Neurodevelopmental Disorders."

14. S. Guandalini et al., "VSL#3 Improves Symptoms in Children with Irritable Bowel Syndrome: A Multicenter, Randomized, Placebo-Controlled, Double-Blind, Crossover Study," *Journal of Pediatric Gastroenterology and Nutrition* 51, no. 1 (July 2010): 24-30.

15. M. Dapoigny et al., "Efficacy and Safety Profile of LCR35 Complete Freeze-Dried Culture in Irritable Bowel Syndrome: A Randomized, Double-Blind Study," *World Journal of Gastroenterology* 18, no. 17 (May 7, 2012): 2067-75.

16. E. Smecuol et al., "Exploratory, Randomized, Double-Blind, Placebo-Controlled Study on the Effects of *Bifidobacterium Infantis* Natren Life Start Strain Super Strain in Active Celiac Disease," *Journal of Clinical Gastroenterology* 47, no. 2 (February 2013): 139-47.

17. M. Frémont et al., "High-Throughput 16S rRNA Gene Sequencing Reveals Alterations of Intestinal Microbiota in Myalgic Encephalomyelitis/Chronic Fatigue Syndrome Patients," *Anaerobe* 22 (August 2013): 50-56.

18. M. Messaoudi et al., "Beneficial Psychological Effects of a

Probiotic Formulation (*Lacto-bacillus Helveticus* R0052 and *Bifidobacterium Longum* R0175) in Healthy Human Volunteers," *Gut Microbes* 2, no. 4 (July/August 2011): 256–61.

SIDEBAR:
A BRIEF HISTORY OF BUGS

1. Medical Council, General Board of Health, *Report of the Committee for Scientific Inquiries in Relation to the Cholera-Epidemic of 1854* (London, England, 1855).

CHAPTER 5:
HACKING YOUR MICROBIOME

1. C. A. Lozupone et al., "Diversity, Stability and Resilience of the Human Gut Microbiota," *Nature* 489, no. 7415 (September 13, 2012): 220–30.

2. S. H. Duncan et al., "Contribution of Acetate to Butyrate Formation by Human Faecal Bacteria," *British Journal of Nutrition* 91 (2004): 915–23.

3. World Gastroenterology Organisation, *World Gastroenterology Organisation Practice Guideline: Probiotics and Prebiotics* (May 2008). www.worldgastroenterology.org/assets/downloads/en/pdf/guidelines/19_probiotics_prebiotics.pdf.

4. To give you a flavor of what the results of clinical trials look like when you read them, in a crossover, placebo-controlled, double-blind study, consuming 30 grams a day of a prebiotic called isomalt (a mixture of the polyols 1-O-∀-D-glucopyranosyl-D-mannitol and 6-O-∀-D-glucopyranosyl-D-sorbitol) for four weeks resulted in a 65 percent increase in the proportion of bifidobacteria and a 47 percent increase in total bifidobacteria cell counts compared with consuming sucrose 5. A. Gostner, "Effect of Isomalt Consumption on Faecal Microflora and Colonic Metabolism in Healthy Volunteers," *British Journal of Nutrition* 95, no. 1 (January 2006): 40–50. In other words, this prebiotic increased the amount of a kind of bacteria often thought to be good, although direct effects on gut function weren't studied. In a different study, where twelve volunteers ingested 10 grams of inulin per day for sixteen days, in comparison with a control period without any supplement intake, *Bifidobacterium adolescentis* increased from 0.89 percent to 3.9 percent of the total microbiota 6. C. Ramirez-Farias et al., "Effect of Inulin on the Human Gut Microbiota: Stimulation of *Bifidobacterium Adolescentis* and *Faecalibacterium Prausnitzii*," *British Journal of Nutrition* 101, no. 4 (February 2009): 541–50.

5. H. Steed et al., "Clinical Trial: The Microbiological and Immunological Effects of Synbiotic Consumption—A Randomized Double-Blind Placebo-Controlled Study in

Active Crohn's Disease," *Alimentary Pharmacology & Therapeutics* 32, no. 7 (October 2010): 872–83.

6. D. Linetzky Waitzberg, "Microbiota Benefits After Inulin and Partially Hydrolized Guar Gum Supplementation: A Randomized Clinical Trial in Constipated Women," *Nutricion* Hospitalaria 27, no. 1 (January/February 2012): 123–29.

7. Z. Asemi et al., "Effects of Synbiotic Food Consumption on Metabolic Status of Diabetic Patients: A Double-Blind Randomized Cross-over Controlled Clinical Trial," *Clinical Nutrition* 33, no. 2 (April 2014): 198–203.

8. J. A. Applegate et al., "Systematic Review of Probiotics for the Treatment of Community-Acquired Acute Diarrhea in Children," supplement 3, *BMC Public Health* 13 (2013): S16.

9. A. P. Hungin et al., "Systematic Review: Probiotics in the Management of Lower Gastrointestinal Symptoms in Clinical Practice—An Evidence-Based International Guide," *Alimentary Pharmacology & Therapeutics* 38, no. 8 (October 2013): 864–86.

10. N. P. McNulty et al., "The Impact of a Consortium of Fermented Milk Strains on the Gut Microbiome of Gnotobiotic Mice and Monozygotic Twins," *Science Translational Medicine* 3, no. 106 (October 26, 2011): 106ra106.

11. H. J. Kim et al., "A Randomized Controlled Trial of a Probiotic Combination VSL# 3 and Placebo in Irritable Bowel Syndrome with Bloating," *Neurogastroenterology and Motility: The Official Journal of the European Gastrointestinal Motility Society* 17, no. 5 (October 2005): 687–96.

12. D. J. Merenstein, J. Foster, and F. D'Amico, "A Randomized Clinical Trial Measuring the Influence of Kefir on Antibiotic-Associated Diarrhea: Measuring the Influence of Kefir (MILK) Study," *Archives of Pediatrics & Adolescent Medicine* 163, no. 8 (August 2009): 750–54; R. S. Beniwal, "A Randomized Trial of Yogurt for Prevention of Antibiotic-Associated Diarrhea," *Digestive Diseases and Sciences* 48, no. 10 (October 2003): 2077–82.

13. "*Clostridium Difficile* Fact Sheet," Centers for Disease Control and Prevention, accessed September 2014, www.cdc.gov /hai/eip/pdf/Cdiff-factsheet.pdf.

14. I. Youngster, "Fecal Microbiota Transplant for Relapsing *Clostridium Difficile* Infection Using a Frozen Inoculum from Unrelated Donors: A Randomized, Open-Label, Controlled Pilot Study," *Clinical Infectious Diseases: An Official Publication of the Infectious Diseases Society of America* 58, no. 11 (June 1, 2014): 1515–22; Z. Kassam et al., "Fecal Microbiota Transplantation for *Clostridium Difficile* Infection: Systematic Review and Meta-Analysis," *American*

Journal of Gastroenterology 108, no. 4 (April 2013): 500–508.

15. "How Well Do Vaccines Work?," Vaccines.gov, US Department of Health and Human Services, accessed October 11, 2014, www .vaccines.gov/basics/effectiveness.

16. "Vaccines · Disease," Immunization Healthcare Branch, accessed October 11, 2014, www .vaccines.mil/Vaccines.

17. Y. Li et al., "On the Origin of Smallpox: Correlating Variola Phylogenics with Historical Smallpox Records," *Proceedings of the National Academy of Sciences of the United States of America* 104, no. 40 (October 2, 2007): 15787–92.

18. Rob Stein, "Should Last Remaining Known Smallpox Virus Die?," *Washington Post*, March 8, 2011.

19. Z. Wang et al., "Gut Flora Metabolism of Phosphatidylcholine Promotes Cardiovascular Disease," *Nature* 472, no. 7341 (April 7, 2011): 57–63.

20. A. D. Kostic et al., "Genomic Analysis Identifies Association of *Fusobacterium* with Colorectal Carcinoma," *Genome Research* 22, no. 2 (February 2012): 292–98.

21. Ley, "Microbial Ecology."

22. C. A. Lowry et al., "Identification of an Immune-Responsive Mesolimbocortical Serotonergic System: Potential Role in Regulation of Emotional Behavior," *Neuroscience* 146, no. 2 (May 11, 2007): 756–72.

23. G. A. Rook, C. L. Raison, and C. A. Lowry, "Can We Vaccinate Against Depression?," *Drug Discovery Today* 17, nos. 9–10 (May 2012): 451–58.

CHAPTER 6:
ANTIBIOTICS

1. "Conjunctivitis (Pink Eye) in Newborns," Centers for Disease Control and Prevention, accessed October 11, 2014, www.cdc.gov /conjunctivitis/newborns.html.

2. J. F. Burns, "British Medical Council Bars Doctors Who Linked Vaccine with Autism," *New York Times*, May 24, 2010.

3. "Possible Side Effects from Vaccines," Centers for Disease Control and Prevention, accessed October 11, 2014, www.cdc.gov /vaccines/vac-gen/side-effects .htm.

4. *New York Times* editorial board, "The Rise of Antibiotic Resistance," *New York Times*, May 10, 2014.

5. M. J. Blaser, *Missing Microbes: How the Overuse of Antibiotics Is Fueling Our Modern Plagues* (New York: Henry Holt, 2014).

6. "Battle of the Bugs: Fighting Antibiotic Resistance," US Food and Drug Administration, last modified August 17, 2011, www .fda.gov/Drugs/ResourcesForYou /Consumers/ucm143568.htm.

7. G. D. Wright, "Mechanisms of Resistance to Antibiotics,"

Current Opinion in Chemical Biology 7, no. 5 (October 2003): 563–69.

8. V. J. Paul et al., "Antibiotics in Microbial Ecology: Isolation and Structure Assignment of Several New Antibacterial Compounds from the Insect-Symbiotic Bacteria *Xenorhabdus* Spp," *Journal of Chemical Ecology* 7, no. 3 (May 1981): 589–97.

9. "Use of Antimicrobials Outside Human Medicine and Resultant Antimicrobial Resistance in Humans," World Health Organization, accessed October 12, 2014, http://web.archive.org/web/20040513120635/http://www.who.int/mediacentre/factsheets/fs268/en/index.html.

10. R. M. Lowe et al., "*Escherichia Coli* O157:H7 Strain Origin, Lineage, and Shiga Toxin 2 Expression Affect Colonization of Cattle," *Applied Environmental Microbiology* 75, no. 15 (August 2009): 5074–81.

11. I. Cho et al., "Antibiotics in Early Life Alter the Murine Colonic Microbiome and Adiposity," *Nature* 488, no. 7413 (August 30, 2012): 621–26.

12. Blaser, *Missing Microbes.*

13. L. Trasande et al., "Infant Antibiotic Exposures and Early-Life Body Mass," *International Journal of Obesity* 37, no. 1 (January 2013): 16–23.

14. S. Foliaki et al., "Antibiotic Use in Infancy and Symptoms of Asthma, Rhinoconjunctivitis, and Eczema in Children 6 and 7 Years Old: International Study of Asthma and Allergies in Childhood Phase III," *Journal of Allergy and Clinical Immunology* 124, no. 5 (November 2009): 982–89.

15. A. T. Stefka et al., "Commensal Bacteria Protect Against Food Allergen Sensitization," *Proceedings of the National Academy of Sciences of the United States of America* 111, no. 36 (September 9, 2014): 13145–50.

ADDENDUM:
THE AMERICAN GUT

1. See American Gut, www.americangut.org.

SIDEBAR:
THE SCIENCE (AND ART) OF
MICROBIOME MAPPING

1. E. S. Lander et al., "Initial Sequencing and Analysis of the Human Genome," *Nature* 409, no. 6822 (February 15, 2001): 860–921.

2. Human Microbiome Project Consortium, "Structure, Function and Diversity."

3. C. Lozupone and R. Knight, "UniFrac: A New Phylogenetic Method for Comparing Microbial Communities," *Applied and Environmental Microbiology* 71, no. 12 (December 2005): 8228–35.

4. Turnbaugh et al., "A Core Gut Microbiome in Obese and Lean

Twins," *Nature* 457, no. 7228 (January 22, 2009): 480–84.

5. Le Chatelier et al., Richness of Human Gut Microbiome Correlates with Metabolic Markers."

6. J. L. Dunne, "The Intestinal Microbiome in Type 1 Diabetes," *Clinical and Experimental Immunology* 177, no. 1 (July 2014): 30–37.

7. F. Soyucen et al., "Differences in the Gut Microbiota of Healthy Children and Those with Type 1 Diabetes," *Pediatrics International: Official Journal of the Japan Pediatric Society* 56, no. 3 (June 2014): 336–43.

8. M. E. Mejia-Leon et al., "Fecal Microbiota Imbalance in Mexican Children with Type 1 Diabetes," *Scientific Reports* 4 (2014): 3814.

9. N. Larsen et al. "Gut Microbiota in Human Adults with Type 2 Diabetes Differs from Non-Diabetic Adults," *PloS One* 5, no. 2 (2010): e9085.

10. F. H. Karlsson et al., "Gut Metagenome in European Women with Normal, Impaired and Diabetic Glucose Control," *Nature* 498, no. 7452 (June 6, 2013): 99–103.

11. J. Sato et al., "Gut Dysbiosis and Detection of 'Live Gut Bacteria' in Blood of Japanese Patients with Type 2 Diabetes," *Diabetes Care* 37, no. 8 (August 2014): 2343–50.

12. J. Qin et al., "A Metagenome-wide Association Study of Gut Microbiota in Type 2 Diabetes,"

Nature 490, no. 7418 (October 4, 2012): 55–60.

13. Frank et al., "Molecular-Phylogenetic Characterization of Microbial Community Imbalances."

14. Tong et al., "A Modular Organization of Human Intestinal Mucosal Microbiota and Its Association with Inflammatory Bowel Disease."

15. E. Li et al., "Inflammatory Bowel Diseases Phenotype, *C. Difficile* and NOD2 Genotype Are Associated with Shifts in Human Ileum Associated Microbial Composition," *PloS One* 7, no. 6 (2012): e26284.

16. D. Gevers et al., "The Treatment-Naive Microbiome in New-Onset Crohn's Disease," *Cell Host & Microbe* 15, no. 3 (March 12, 2014): 382–92.

17. C. Manichanh et al., "Anal Gas Evacuation and Colonic Microbiota in Patients with Flatulence: Effect of Diet," *Gut* 63, no. 3 (March 13, 2014): 401–8.

18. Castellarin et al., "*Fusobacterium Nucleatum* Infection Is Prevalent in Human Colorectal Carcinoma."

19. Rubinstein et al., "*Fusobacterium Nucleatum* Promotes Colorectal Carcinogenesis by Modulating E-Cadherin/Beta-Catenin Signaling via Its FadA Adhesin."

20. Kostic et al., "*Fusobacterium Nucleatum* Potentiates Intestinal Tumorigenesis and

Modulates Tumor-Immune
Microenvironment."

21. Warren et al., "Co-occurrence of
Anaerobic Bacteria in Colorectal
Carcinomas."

22. Flanagan et al., "*Fusobacterium
Nucleatum* Associates with
Stages of Colorectal Neoplasia
Development, Colorectal Cancer
and Disease Outcome," 1381–90.

23. Koeth et al., "Intestinal Microbiota Metabolism of L-Carnitine."

24. Tang et al., "Intestinal Microbial
Metabolism of Phosphatidylcholine and Cardiovascular Risk."

25. Scher et al., "Expansion of Intestinal *Prevotella Copri* Correlates
with Enhanced Susceptibility to
Arthritis."

26. F. H. Karlsson et al., "Gut Metagenome in European Women
With Normal, Impaired and
Diabetic Glucose Control," *Nature* 498, no. 7452 (June 6, 2013):
99–103.

27. J. Qin et al., "A Metagenome-Wide
Association Study of Gut Microbiota in Type 2 Diabetes," *Nature*
490, no. 7418 (Oct. 4, 2012): 55–60.

28. D. Knights et al., "Human-
Associated Microbial Signatures:
Examining Their Predictive
Value," *Cell Host & Microbe* 10,
no. 4 (Oct. 20, 2011): 292–6.

29. D. Gevers et al., "The Treat-
ment-Naive Microbiome in
New-Onset Crohn's Disease,"
Cell Host & Microbe 15, no. 3
(March 12, 2014): 382–92.

30. H. S. Lee et al., "Biomarker
Discovery Study Design for Type 1
Diabetes in the Environmental
Determinants of Diabetes in the
Young (TEDDY) Study," *Diabetes/
Metabolism Research and Reviews*
30, no. 5 (July 2014): 424–34.

31. Lee et al., "Proinflammatory
T-cell Responses to Gut Microbiota Promote Experimental Autoimmune Encephalomyelitis."

32. Berer et al., "Commensal
Microbiota and Myelin Auto-
antigen Cooperate to Trigger
Autoimmune Demyelination."

33. Hsiao et al., "Microbiota Modulate Behavioral and Physiological Abnormalities Associated
with Neurodevelopmental
Disorders."

34. C. A. Lozupone et al., "Meta-
analyses of Studies of the
Human Microbiota," *Genome
Research* 23, no. 10 (October
2013): 1704–14.

35. Hamady and Knight, "Microbial
Community Profiling for Human
Microbiome Projects."

36. Z. Liu et al., "Accurate Tax-
onomy Assignments from
16S rRNA Sequences Produced
by Highly Parallel Pyrosequenc-
ers," *Nucleic Acids Research* 36,
no. 18 (October 2008): e120.

37. Z. Liu et al., "Short Pyro-
sequencing Reads Suffice for
Accurate Microbial Commu-
nity Analysis," *Nucleic Acids
Research* 35, no. 18 (September
2007): e120.

ABOUT THE AUTHORS

ROB KNIGHT is a professor of pediatrics, and computer science and engineering, and Director of the Microbiome Initiative at the University of California, San Diego. He is cofounder of the American Gut Project and the Earth Microbiome Project.

BRENDAN BUHLER is an award-winning science writer whose work has appeared in the *Los Angeles Times, California,* and *Sierra Magazine.* His story on Rob Knight's work was selected for the 2012 edition of *The Best American Science and Nature Writing.*

WATCH ROB KNIGHT'S TED TALK

Rob Knight, author of *Follow Your Gut*, spoke at the TED Conference in 2014. His talk, available for free at TED.com, was the inspiration for this book.

PHOTO: JAMES DUNCAN DAVIDSON/TED

RELATED TALKS ON TED.COM

Jessica Green
Our bodies are covered in microbes. Let's design for that.

Our bodies and homes are covered in microbes—some good for us, some bad for us. As we learn more about the germs and microbes who share our living spaces, TED Fellow Jessica Green asks: Can we design buildings that encourage happy, healthy microbial environments?

Bonnie Bassler
How bacteria talk

Bonnie Bassler discovered that bacteria "talk" to one another, using a chemical language that lets them coordinate defense and mount attacks. The finding has stunning implications for medicine, industry—and our understanding of ourselves.

Ed Yong
Suicidal crickets, zombie roaches, and other parasite tales

We humans set a premium on our own free will and independence, and yet there's a shadowy influence we might not be considering. Parasites have perfected the art of manipulation to an incredible degree. So are they influencing us? It's more than likely.

Jonathan Eisen
Meet your microbes

Our bodies are covered in a sea of microbes—both the pathogens that make us sick and the "good" microbes, about which we know less, that might be keeping us healthy. At TEDMED, microbiologist Jonathan Eisen shares what we know, including some surprising ways to put those good microbes to work.